Enterprising Nature

Antipode Book Series

Series Editors: Vinay Gidwani, University of Minnesota, USA and Sharad Chari, CISA at the University of the Witwatersrand, South Africa

Like its parent journal, the Antipode Book Series reflects distinctive new developments in radical geography. It publishes books in a variety of formats – from reference books to works of broad explication to titles that develop and extend the scholarly research base – but the commitment is always the same: to contribute to the praxis of a new and more just society.

Published

Enterprising Nature

*Economics, Markets, and Finance
in Global Biodiversity Politics*

Jessica Dempsey

WILEY Blackwell

This edition first published 2016
© 2016 John Wiley & Sons, Ltd.

Registered Office
John Wiley & Sons, Ltd, The Atrium, Southern Gate, Chichester, West Sussex, PO19 8SQ, UK

Editorial Offices
350 Main Street, Malden, MA 02148-5020, USA
9600 Garsington Road, Oxford, OX4 2DQ, UK
The Atrium, Southern Gate, Chichester, West Sussex, PO19 8SQ, UK

For details of our global editorial offices, for customer services, and for information about how to apply for permission to reuse the copyright material in this book please see our website at www.wiley.com/wiley-blackwell.

Library of Congress Cataloging-in-Publication Data

Names: Dempsey, Jessica, 1977– author.
Title: Enterprising nature : economics, markets, and finance in global
  biodiversity politics / Jessica Dempsey.
Description: Chichester, UK ; Hoboken, NJ : John Wiley & Sons, 2016. |
  Includes bibliographical references and index.
Identifiers: LCCN 2016010099 | ISBN 9781118640609 (cloth) | ISBN 9781118640555
  (pbk.) | ISBN 9781118640531 (ePub) | ISBN 9781118640548 (Adobe PDF)
Subjects: LCSH: Environmental economics. | Environmental policy–Economic aspects.
Classification: LCC HC79.E5 D45155 2016 | DDC 333.7–dc23 LC record available at
  https://lccn.loc.gov/2016010099

A catalogue record for this book is available from the British Library.

Cover image: Illustration by Ian Whadcock / ianwhadcock.com.
Cover design by Wiley and Cathy Matusicky, State Creative Group

Set in 10.5/12.5pt Sabon by SPi Global, Pondicherry, India

1   2016

Contents

Acronyms

AHTEG	Ad Hoc Technical Expert Group
ALBA	Bolivarian Alliance for the Peoples of Our America
BAU	Business as Usual
BBRs	Biodiversity Business Risks
BP	British Petroleum
BSR Network	Business Social Responsibility Network
CBD	Convention on Biological Diversity
CDM	Clean Development Mechanism
CEQ	Council on Environmental Quality
CER	Certified Emission Reductions
CESR	Corporate Ecosystem Service Review
CEV	Corporate Ecosystem Valuation
CI	Conservation International
COP	Conference of the Parties
COP 9	9th Conference of the Parties to the Convention on Biological Diversity
COP 10	10th Conference of the Parties to the Convention on Biological Diversity
CRP	Conference Room Paper
CSD	Commission on Sustainable Development
CSR	corporate social responsibility
EIRIS	Experts in Responsible Investment Solutions
ES	Ecosystem Services
ETS	European Trading Scheme
FAO	Food and Agriculture Organization
FI	Financial Institution
FFI	Flora & Fauna International
G77	Group of 77
GBA	Global Biodiversity Assessment

GDM	Green Development Mechanism
GDP	gross domestic product
GE	Green Economy
GEF	Global Environment Facility
GMO	genetically modified organism
HoB	Heart of Borneo
IBAT	Integrated Biodiversity Assessment Tool
ICSF	International Collective in Support of Fishworkers
IFC	International Finance Corporation
IFM	Innovative Financial Mechanisms
IIFB	International Indigenous Forum on Biodiversity
INBio	Costa Rica's National Biodiversity Institute
InVEST	Integrated Valuation of Ecosystem Services and Trade-offs
IPCC	Intergovernmental Panel on Climate Change
IRR	internal rate of return
IUCN	International Union for the Conservation of Nature
MA	Millennium Ecosystem Assessment
MSY	Maximum Sustained Yield
NAS	National Academy of Sciences
NatCap	Natural Capital Project
NBASP	National Biodiversity Action and Strategies Plan
NCD	Natural Capital Declaration
NGO	Non-governmental organization
PES	Payments for Ecosystem Services
PwC	PricewaterhouseCoopers
PWS	payments for watershed services
REDD	Reducing Emissions from Deforestation and forest Degradation
SAR	species–area relationship
SBSTTA	Subsidiary Body on Scientific, Technical and Technological Advice
SCOPE	Scientific Committee on the Problems of the Environment
TEEB	The Economics of Ecosystems and Biodiversity project
TNC	The Nature Conservancy
TWN	Third World Network
UBC	University of British Columbia
UN	United Nations
UNDP	United Nations Development Programme
UNEP	United Nations Environment Programme
UNEP FI	United Nations Environment Programme Finance Initiative

UNEP-WCMC	United Nations Environment Programme World Conservation Monitoring Centre
UNESCO	United Nations Educational, Scientific and Cultural Organization
UNFCCC	United Nations Framework Convention on Climate Change
USAID	US Agency for International Development
WB	World Bank
WBCSD	World Business Council on Sustainable Development
WCED	World Commission on Environment and Development
WEF	World Economic Forum
WGRI	Working Group on the Review of Implementation of the Convention on Biological Diversity
WRI	World Resources Institute
WTO	World Trade Organization
WWF	World Wildlife Fund

Series Editors' Preface

The *Antipode Book Series* explores radical geography "antipodally," in opposition, from various margins, limits, or borderlands.

Antipode books provide insight "from elsewhere," across boundaries rarely transgressed, with internationalist ambition and located insight; they diagnose grounded critique emerging from particular contradictory social relations in order to sharpen the stakes and broaden public awareness. An *Antipode* book might revise scholarly debates by pushing at disciplinary boundaries, or by showing what happens to a problem as it moves or changes. It might investigate entanglements of power and struggle in particular sites, but with lessons that travel with surprising echoes elsewhere.

Antipode books will be theoretically bold and empirically rich, written in lively, accessible prose that does not sacrifice clarity at the altar of sophistication. We seek books from within and beyond the discipline of geography that deploy geographical critique in order to understand and transform our fractured world.

Vinay Gidwani
University of Minnesota, USA

Sharad Chari
CISA at the University of the Witwatersrand, South Africa

Antipode Book Series Editors

Preface

The first spark of this book began in May 2006, in the outskirts of Curitiba, Brazil. I was attending a negotiation of the United Nations Convention on Biological Diversity (CBD). Most attention centered whether or not the Parties, government signatories to the agreement, would reaffirm or overturn a moratorium on the field-testing of what is known colloquially as terminator technology (seeds engineered to produce sterile seeds). Hundreds of small farmers and landless people gathered outside the conference center every day reminding the suited delegates that they had responsibilities beyond the patent holders of the technology.

Outside this crucial debate, other agendas galloped ahead. The first of many events on concepts like "biodiversity offsets" took place, and bureaucrats were just beginning to speak in the language of ecosystem services. Compared to other CBD negotiations, where debates oriented around the definition of "primary forests," it seemed as though the floor underneath international conservation was shifting. Global biodiversity policy was going (more) economic, and perhaps market based! This research was conceived following that negotiation, oriented around a simple question: how did this happen? How did economic and market-based approaches become so dominant, even commonsense, in global biodiversity conservation?

My role in these biodiversity circuits has never been one of passive observer, but of active participant, largely organized by the ongoing work of the Convention on Biological Diversity Alliance (CBD Alliance), a network of civil society groups that follows and intervenes in global biodiversity policies. For over a decade, working with all kinds of people, from all kinds of organizations and social movements – from WWF to Via Campesina – I researched and prepared briefing papers, coordinated joint policy statements, and fundraised endlessly to bring Southern NGOs,

Indigenous communities, and social movement representatives to negotiations. Attending over a dozen negotiations, we worked to influence the
shape of international biodiversity law and policy. This might sound as if
I inhabited a glamorous world of international diplomats and the jet-set
crowd, but I can say that it mostly involved sitting with headsets on for
long periods of time, carefully following boring legalese as it shifted and
shaped, crafting alternative text to circulate to friendly government delegates, and working all hours for one or two weeks.

These experiences, but especially the people I worked closely with,
contributed to the particular lens through which I see "enterprising
nature," a phrase that I use to describe efforts to transform diverse
natures into economically competitive entities. More than anything, I
learned how to inhabit the uncomfortable, impure spaces of liberal
environmentalism.

Let me explain. I went to my first CBD negotiation in Den Haag in
2002 armed with a straightforward narrative about the limits of the
global, and especially the limits of the multilateral, a lens honed over
the course of my undergraduate education and local political activities:
the bad experts and elites of the globe continue to wreak havoc on the
local, the Indigenous, the peasant, even when they are saving nature. Yet
upon arrival I met a group of international activists, such as Ricardo
Carrere, Pat Mooney, Ashish Kothari, Chee Yoke Ling, Patrick Mulvany,
and Simone Lovera, who were at once deeply skeptical of the premises
of the CBD and the "sustainable development compromise," but who
also used the negotiations as a site to draw attention to the persistent
blind spots in international environmental law and policy: to how new
financial mechanisms fail to address deep power imbalances and socioecological injustices, to the way that very small steps forward at the CBD
are undermined by neoliberal trade rules, to the enormous gulfs between
haves and have-nots, to the epistemological conceits of Western
conservation and science.

They were (and still are) constantly reminding government delegates
and international experts that global biodiversity loss is an effect of a
kind of "imperial ruination" (to take a term from Anne Stoler), and that
addressing this problem requires not just cooperation and consensus
between nations, but also disassembling deeply etched power asymmetries and clusters of concentrated power and knowledge that mark some
ways of knowing, valuing, and living with nature above others – over
and over again, with violent effects for both humans and nonhumans.
This tireless group of people showed me what global environmental justice politics looked like: the problems of trenchant poverty in the Global
South and the sixth extinction were not oppositional problems, but
rather problems with the same root.

But perhaps most crucially, I learned that there was no privileged or perfect place to conduct this struggle; on one day I was holding a banner outside the negotiation stating "no green economy," then a bit later I was circulating concrete language to improve the Convention text, asking for further study and research on the impact of market-based approaches. We were engaged in advocacy that sometimes shamed governments, sometimes destabilized the worst policy initiatives, and occasionally saw victories (as with the ban on terminator technology that was reaffirmed by governments in Curitiba).

The research questions and approach of this book are overdetermined by this set of personal and political experiences: I see the global as neither homogenous nor smooth, but replete with contestation and even possibility. The shifting ground toward "enterprising nature" is deeply inflected by the hegemonic, elite processes of contemporary neoliberal capitalism, but it is also composed of people I know, often found easy to talk to, and with whom I could at times imagine becoming allies, depending on the issue or the political moment. And over the course of my research I was often surprised at what the most ardent advocates of "enterprising" said in the course of interviews, at the difficulties and hesitations articulated.

...

In this book I tell the story of how biodiversity is being tethered to economic and market logics and practices: when and where this is happening, who and what is involved, and how it is unfolding. This is the story of the making of enterprising nature. But the book is also a story of its *non-making*: attempts "to enterprise" are often halting and even marginal (while remaining strangely hegemonic).

Within this book I do not dismiss the people involved in enterprising nature, or their ideas, their knowledges and tools. My aim is to open a historically and geographically situated debate on this way of addressing the monoculturing of the planet. The approach I bring to enterprising nature is influenced by feminist scholarship, particularly that of Donna Haraway. Her work reminds us that science and technology are accumulation strategies, deeply implicated in producing classed, gendered, and racialized hierarchies. Yet she also asks that we sit awhile with the excess, with historical and geographical overabundance, that we engage with the "always messy projects of description, narration, intervention, inhabiting, conversing, exchanging, and building" (1994, 62). The point is not simply to "make a tangled mess," but rather to "learn something about how worlds get made and unmade, and for whom" (1994, 70).

My hope is that this book will be of interest to scholars engaged with debates over the character of environmentalism and conservation in an era of neoliberalism. I hope, too, that it will be read by actors in the circuits I describe: scientists, economists, bureaucrats, employees of non-profit and international organizations, who are keen to reflect on broader implications of the processes in which they are enmeshed. While enterprising nature is so pervasive as to seem axiomatic, if we look closely at the specific operations of these circuits and calculative devices, their effects and non-effects, we can avoid the weary resignation, narrowing vision, and sense of inevitability that so many involved in resisting the sixth extinction nowadays experience. For this reason, I am particularly hopeful that activists and advocates in biodiversity politics will see the book as an invitation to engage and challenge the turn to enterprising nature in new ways.

...

I remember when I was a graduate student I would read other people's book acknowledgments and think: why did it take them so long? Now I get it. This book took a stupidly long time to finish and so there are many people to thank.

For the most part this project emerged out of an almost decade-long participation in two worlds – the University of British Columbia's Department of Geography where I was a student and around the world of global biodiversity politics. Around the negotiations of the Convention on Biological Diversity, I can't believe my luck in meeting the inspiring, insightful and fierce policy wonks and activists already mentioned above, and many more, such as Faris Ahmed, Tasneem Balasinorwala, Joji Carino, S. Faizi, Ana Filipini, Barbara Gemmill, Antje Lorch, Malia Nobrega, Helena Paul, Hope Shand, Chandrika Sharma, Ricarda Steinbrecher, Jim Thomas, and Christine von Weizsacker. All of these people infuse this research project, although they may take issue with some of my interpretations.

Many other people generously gave of their time to be interviewed in the course of my research. It was rare that anyone said no, strangely, even if it meant sneaking me into a 15-minute time slot (which often carried on much longer). The Trudeau Foundation and the Social Sciences and Humanities Research Council supported this travel-intensive research.

At UBC, Trevor Barnes provided the most supportive but also demanding supervision on my dissertation that led to this book. I don't want to make anyone jealous, but he's the kind of supervisor who turns your work around to you overnight (it's true, it happened many times!) but also makes your heart skip a beat because you know that feedback is going to feel like ripping a band-aid off raw skin. Also at UBC, Gerry Pratt and Juanita Sundberg led me through rigorous scholarly but also political debates in courses, over beer and in the hallways at 1984 West Mall. They model a kind of feminist, political scholarship that I aspire to: keenly critical and questioning but also open and generous. I'm so happy to be back there as their colleague.

If there is anything good in this book it is in good part due to the diligent advising I received at UBC but also the ever-critical and supportive
students and postdocs I met there: Chris, Matt S., Bonnie, Ted, Alex,
Pablo, Fiona, Tyler, Sarah, Michael, Joel, Dawn, and many more. Kevin
Gould, Matt Dyce, Shiri Pasternak, Jono Peyton, Jo Reid, Geoff Mann,
Emilie Cameron, and Rosemary Collard are probably some of the funni-
est and most fun people to be a student and now faculty with, as well
as hefting serious scholarly weight. I am so grateful for their ongoing
friendship and collaborations. Emilie, Rosemary, Geoff, and Jo in
particular are big influences on this book: Emilie with her straight-
shooting advice and wise insight over the phone, Geoff with his political-
economic might delivered from his perch at JJ Bean (and often behind a
can of 1516), Rosemary with her lightning-fast email responses clari-
fying and improving anything I put to paper. Jo edited and re-edited the
manuscript, raising big and little questions about what lies within. They
are the jackpot I somehow managed to hit in this whole academic thing.

Others who improved this book with their astute readings or conversations include Scott Prudham, Karen Bakker, Matthew Sparke, Nik
Heynen, Laura Janara, Rebecca Lave, Raj Patel, Juliane Collard, Erika
Bland, Ryan Lucy, Morgan Robertson, Larry Lohmann, Daniel Suarez,
and Sian Sullivan. Then there is the stellar editing of Vinay Gidwani, all
the way from book prospectus to final edits. I'm so happy I was on the
receiving end of his hard questions, thoughtful comments, and ongoing
support. At Wiley-Blackwell, Jacqueline Scott and Sakthivel Kandaswamy
were ever so patient and prompt; copy-editor Katherine Carr whipped
the text into further shape. Cathy Matusicky designed the lovely cover;
Ian Whaddel allowed us to use his illustration and Eric Leinberger at
UBC helped with maps.

Much of this book was rewritten (and re-rewritten) while I was at the
University of Victoria's School of Environmental Studies, an oasis in the
dry, ever neoliberalizing university. Conversations with Kara Shaw,
James Rowe, Michael M'Gonigle, and Brian Starzomski are highlights
of my time there, and they read parts of this book with their keen analytical but also political minds. Michael also sent me to my first
Convention on Biological Diversity meeting in 2002, setting off the more
than a decade of learning and collaboration that underpins this book.
James and Kara deserve special fist bumps for having my back in those
sometimes challenging years.

All the way through are friends and family, many of whom are the
pointy-heads mentioned above, but also Suzie, Shawn, Narda, Deb,
Brad, Michelle, Kira, Donovan, Madeline, Robbie, Nate, Trish and all
of the little people that surround us, crack us up, and irritate us.
My mom and dad – Joy and Steve – regularly dropped everything to

drive from Edmonton to Victoria/Vancouver to help out, providing
bucketloads of support and love to my family when I was travelling.
There is also our extended family of caregivers for my kids for over a
decade: Lynn Busby, Laurel Beerbower, Narda Nelson, Eaglets, the
YMCA-False Creek and even further back. My now not-so-little
family is sustained by a big circle.

In 2006 when I went to Curitiba for the CBD negotiation, my family
came along for the ride – then composed of Ryan and Sean. Sean was not
quite 2. Now he is 11. I sent off the first book prospectus for this project
in 2012 when I was nearly exploding with who are now Cecelia and
Eloise, the irrepressible CC and Elly. And so the biggest high five and the
most explosive fist bump must go to the ever-generous, wise, and
(mostly) unflappable Ryan, the key condition of possibility for all this
life and liveliness. Thank you, my love!

Vancouver, Canada
February 2016

1

Enterprising Nature

In the Beginning, There Was Failure

In book- and paper-stuffed academic offices, walking down cold and dark streets in Norway alongside government bureaucrats, on Skype interviews with bankers – everywhere I went in the course of my research people talked about the failures of biodiversity conservation. "We tried to make people care about nature for its own sake," said global experts, "without the results." I read about failure within the pages of *Science* and *Nature*; I decoded profound disappointment in the stilted text of multilateral policy documents. Over beer in a noisy Palo Alto bar, the chief scientist of The Nature Conservancy, Peter Kareiva, explained the problem in his straight-shooting manner, "No one cares about biodiversity outside of the Birkenstock crowd." Biodiversity, he went on to say, "is something that suburban white kids care about and nobody else."

While I remain unconvinced that no one cares about biodiversity outside of white, suburban hippies, such tired resignation makes sense. The decimation of nonhuman life on earth continues. Despite conservation-oriented laws and policies at every level of governance from local to international, and the establishment of thousands of protected areas, "there is no indication of a significant reduction in the rate of biodiversity loss, nor of a significant reduction in pressures upon it" (CBD 2010a, 17). A study published in *Science* found that most indicators of the state of biodiversity are in decline, and the pressures underlying this shift are also increasing. One in five species of vertebrates are classified as threatened, with that figure increasing every year; 322 vertebrate species have gone extinct since 1500. Declining diversity is apparent in agriculture, where 75% of genetic diversity has

Enterprising Nature: Economics, Markets, and Finance in Global Biodiversity Politics,
First Edition. Jessica Dempsey.
© 2016 John Wiley & Sons, Ltd. Published 2016 by John Wiley & Sons, Ltd.

been lost since the 1900s. Marine ecosystems, too, face mounting pressures; one quarter of oceanic pelagic sharks and rays are classified as threatened or near threatened. What threatened status indicates is a loss of overall abundance of most animals almost everywhere on the earth, a process biologist Rodolfo Dirzo and his colleagues term defaunation: they estimate there are 28% fewer vertebrate animals across species today than there were only four short decades ago; a startling 35% fewer butterflies and moths over the same time period.[1] In a conference hall in Trondheim, Norway, Robert Watson, former chief scientist at the World Bank, declared that 2010 – the UN-declared Year of Biological Diversity – should be a time not to celebrate, but rather to mourn biodiversity's loss.

What does this loss spell for the future of the planet? The impacts of biodiversity loss on global ecosystem function are difficult to study, and even more difficult to pin down with any certainty. In spite of focused research programs around the world, scientific understanding of the functional role of biodiversity remains in many ways elusive. Certainly there are risks to living on a planet that's less diverse, but those risks remain hard to quantify in general terms. What these alarming statistics tell us is that we are living in a world that is becoming distinctively less lively, less colorful, and less diverse in the realm of nonhuman life. A key assumption I make in this book is that this earth simply *is* a better place with more color, more kinds of lives, and more ways of living and living with nonhumans. The radical difference of which biodiversity is a part – what many now call biocultural diversity – *matters*.

What can be done to stem this tide of loss? How can beleaguered environmental activists, bureaucrats, and ecologists generate the political will to spur governments, business, and the general population to take the urgent action that's needed? For many ecologists and their allies, the answer lies in a turn toward economics. "The majority of the global population now lives in cities and is disconnected from nature," said Pavan Sukhdev, the head of a major international initiative to economically value biodiversity. This is "not just a physical distance but also an emotional distance." This disconnection, Sukhdev went on, is "so real" that "we have got to speak the language of economics to show there is a connection." For many actors concerned with the conservation of biodiversity, a turn to economics feels like the last hope. Biodiversity, a jaded Canadian bureaucrat-scientist explained to me, must be made relevant to the Ministry of Finance for it to survive. For world-renowned biologist Hal Mooney, there is an urgent need to turn biodiversity into something that both policy-makers and citizens can care about. "When you say that biodiversity delivers services which are a benefit to society," Mooney told me, you begin to speak in language policy-makers understand, and they can "go to their constituents and say, biodiversity is really important for you, personally, because of the services it provides."

In this book, I explore this turn to economics, the efforts to speak a new language in global biodiversity conservation. *Enterprising Nature* is a critical exploration of the ascent of what is becoming a new maxim in this field: "In order to make live, one must make economic."[2] In other words, for diverse nonhumans to persist, biodiversity conservation must become an economically rational policy trajectory, sometimes even profitable. The proliferation of this mantra is the analytical target of this book, which investigates the roots of this refrain and the international alliances and relations that cohere in producing it.

Drawing on four years of intensive, multi-site field research in places such as Nairobi, London, and Nagoya, and on my decade-long involvement in global biodiversity policy-making, this book traces disciplinary apparatuses, ecological-economic methodologies, computer models, business alliances, and regulatory conditions that, together, I argue, aim to create the conditions wherein nature, or parts of nature, can prove itself as "enterprising." This is a nature that no longer needs the bonds of human care or ethical concern, a nature that is certainly not a public investment burden. Rather, the hope is this will be a nature that is entrepreneurial, a nature that can compete not only in the marketplace but also in modern state governance. An *enterprising nature*.

Enterprising nature seems, theoretically and practically, an approach to biodiversity conservation that is entirely compatible with current, predominantly capitalist, global political-economic relations. Producing enterprising nature, however, as this book chronicles, is not straightforward or easy. Challenges arise at every step: there are scientific debates over how biodiversity supports ecological functions and services, and methodological debates on how to tether ecological data to economic value. Also prominent are geopolitical struggles and global political economic forces that have hampered international conservation for decades. The result is that this increasingly dominant discourse remains, by and large, on the margins of policy-making and capital flows.

The story of enterprising nature, then, holds an alarming paradox. Conservation is trying to make itself more relevant to market and state governance through economization, but all these efforts fail to become operational in a way that can let diverse ecologies live. Enterprising nature, I argue, is best conceptualized as promissory, a socioecological-economic utopia whose realization is always *just around the corner*. The story of enterprising nature is one of waiting, of waiting for the conditions that can make the work of nonhumans legible to processes of liberal governance and perhaps facilitate their entry into mainstream processes of accumulation.

Are You Being Served? Two Images of Enterprising Nature

An image from a 2005 edition of the *Economist* reveals the persistent tensions in the enterprising nature ideal. That image, which appears as this book's cover, shows a sharp-looking, somewhat jolly, white, middle-aged accountant behind a desk, doling out money to an orderly line of half-human, half-plant/animal creatures, which appear as happy-ish and perhaps bored laborers. It's payday in a tropical location of some sort and the creature's hand movements suggest impatience. All recipients of the bags of money defy the human–nonhuman species boundary in some ways; a half-conifer–half-man is followed by a mountain goat–sheep-with-boots, a hand-bag-toting, high heels–wearing bald eagle, followed by an odd-looking leopard or maybe jaguar. The image reveals the dream of enterprising nature: orderly, efficient socioecological relations mediated through a monetary transaction.

In the *Economist*, the cartoon accompanied an article titled "Are You Being Served?" that followed the release of the Millennium Ecosystem Assessment (MA). The MA was the first global survey of ecosystem services, a study warning that changing ecologies are increasingly impacting human well-being. Despite the dire findings of this assessment, however, the tone of the *Economist* article is enormously optimistic. The article heralds a new age of ecological-economic accounting. While ecosystem valuation, the *Economist* staff writer notes, was at one time "a fraught process," it is now "improved," mostly due to the knowledge of ecologists, who "know a great deal more than they used to about how ecosystems work" (*Economist* 2005, 77): they know how different ecosystems deliver services (such as water purification, fiber production, carbon sequestration, etc.) and in what quantities. We know, in other words, how we are "being served."

The tone of the article suggests not simply improvement in knowledge and accounting, but the arrival at a milestone in human progress: diverse creatures and systems can now be fully brought into the balance sheets of firms and governments, informing the most efficient state or firm investments or payments. "There is no longer any excuse for considering them [ecosystems] unquantifiable," the article reports (77). The optimism of the *Economist* article – in spite of the fact that it is describing a devastating report of a planet being rendered less and less hospitable to humans – lay within the certainty that an ecological-economic synthesis (as found in the MA) will tell us, once and for all, *how to live* on this planet, how to create a permanent order of socioecological relations. In its depiction of an orderly line-up of creatures awaiting payment, the cartoon shows a triumph of a rational system of value allocation, of enterprising nature. The image, though, sidelines critical questions at the heart of this project: what is the right way for humans to live in relation to nonhuman nature? Who decides this and from what location? What socioecologies are investable, worthy of payment, and on whose terms and authority?[3]

The desire for enterprising nature is a powerful, end-of-history call that reveals troubling signals in environmental politics. What this "will to enterprise" shows, I argue, is a desire for a neutral, objective, efficient, and automatic relationship with nonhuman bodies and populations. Nonhuman nature, this increasingly influential approach pronounces, is best inhabited via an *accounting relationship*, one that can tell us, neutrally and objectively, through an ecological-economic calculus, how to optimize allocations of the services nonhumans provide. This way of thinking about the human place in nature reasserts, perhaps more than ever, a will for human omniscience regarding relations among humans and nonhumans, a "god trick" here dangerously articulated with neoclassical economics.[4] Though an economic logic may show the need for greater investments in nonhuman lives, that investment must be efficient and selective; it is not for all.

That not everything can be saved poses yet another paradox for this kind of conservation. Enterprising nature instrumentalizes, subjugates, and attempts to make *passive* its subjects – diverse living things – in order to save them.[5] It makes new hierarchies of life, rankings meant to guide governance processes. But by rationalizing and, in many ways, *rationing* nonhuman life through accounting and the logic of efficiency, those advocating for an enterprised nature seem to be trying to *lessen* human domination over nonhuman living beings, at least in one register: they desire to reduce ecological impoverishment, to make more space for other species to live on this planet. As Stanford biologist Hal Mooney explained to me, ecosystem service and economic valuation are for many conservationists a means to an end: conserving biodiversity. Enterprising nature then, is fraught with a broader tension: it has an end goal of lessening human impact or even domination of nature, yet aims to achieve it through increasingly instrumentalized knowledge-power frameworks and practices that seem to increase human domination over the nonhuman.[6] Further, enterprising nature appears, from the cover image, as a reassertion of a *particular* human mastery, reproducing familiar geopolitical and racial orders: the white man doling out payments for socioecological relations or beings that serve his interests.

A second image: Buying the axe

In the same *Economist* article about nature's services, a second image shows a suited, happy man towering over an alarmed *campesino* while taking away his axe (see Figure 1.1). In his other hand, the businessman is holding a big bag of money over the *campesino's* head; coercion and

Figure 1.1   Illustration from *The Economist,* © Ian Whadcock (by permission).
First appeared in *The Economist,* 2005.

cash payments appear intimately linked. The image portrays an imbal-
ance of power and continuity with colonial relations but, like colonial
relations, is also ambiguous. Who is the man holding the bag of cash,
and what does he seek to achieve? Does he represent a firm producing
forest carbon credits for profit, continuing colonial circuits of extractive wealth production that grab land to benefit Northern fat-cat
capitalists? The man's smile and grandfatherly sweater vest suggest a
polite transaction from the point of view of the man, perhaps with
honorable, not-entirely-capitalist-intentions. Maybe the man represents an environmental organization like Conservation International
and he is buying the land or the forest concession to preserve a
"biodiversity hotspot"? Or perhaps the man represents a Norwegian

bureaucrat, offering environment-development aid to say, Brazil, in exchange for keeping forests standing and thus doing the critical work of carbon sequestration.

Conservationists, bureaucrats, scientists, and diplomats in favor of enterprising nature are not necessarily animated by material interests; their intentions can be deeply benevolent, animated by liberal rationales of improvement. They are often more akin to what Tania Li (2007) calls "trustees," experts seeking the proper, "right," and perhaps optimal management of "men and things" (Foucault 1991, 93). Development, for Li, remains a deeply colonial project animated by a liberal "will to improve" the lives of subjects. Perhaps then, we might view the man with the bag of cash as a renovated form of the white, Western "savior," à la Rudyard Kipling's "White man's burden." Renovated because the altered conduct – the wresting of the axe – is achieved through a monetized economic transactions that is ostensibly "fair." Fair for the sweater-vested man who achieves conservation or maybe carbon sequestration in return for his bag of cash. And also fair, so the image ambiguously hints, for the *campesino* who receives a bag of money equal in circumference to the tree stump at his feet. In short, the fairness of exchange value: a tree for its market price. What is on offer, then, is a promise of mutually beneficial improvement through a purportedly neutral ecological-economic calculation.

The image reminds us, however, that even benevolent actors seeking "neutral improvement" or "rational" land management are still deeply embedded in colonial circuitry and power relations. Recognition of benevolence by no means vindicates colonialism; rather, as scholars like Li, Domenico Losurdo, and Lisa Lowe remind, the growth of liberal ideals in governance, economics, and culture "have been commensurate with, and deeply implicated in colonialism, slavery, capitalism, and empire" (Lowe 2015, 2).[7] And, depending on the specific time and place, liberal ideals are often tied up in material interests.[8] Biodiversity has always been deeply liberal in this sense, entangled in universal impulses for the good of all humanity as well as economistic and imperialist drivers (see chapter 2).

Indeed, the image reflects a well-worn trope in international environmentalism – poor people like this *campesino* cause biodiversity loss through demand for firewood or income, problems to be solved in this case by paying him. Critical scholarship by the likes of Vandana Shiva (1991) and Arturo Escobar (1998) point, again and again, to how representations of individuals and communities in the Global South dehistoricize and depoliticize the colonial and imperial institutions and impositions that so often lead to dispossession, displacement, and biodiversity loss. This is a key point made by decades of political ecological research: global environmental policies and politics too often direct our attention away from World Bank loans that fuel monocultures and green revolutions, preferring to focus our attention on the poor person with the axe. A trustee tends to ask how one might improve the situation by enhancing the ability of the individual to "do good" (here the peasant with the axe), perhaps via an economic incentive: a tactic that obscures the messy politics and social relations that produce deforestation and the *campesino*'s involvement in it.

Along these lines, the image in the *Economist* accompanies discussion of a proposed forest bond scheme. In the article, John Forgach, a principal of ForestRe, a forestry insurance company based in London, describes his idea to create 25-year "forest-backed bonds" that would fund a massive reforestation alongside the Panama Canal in hope of reducing siltation and maintaining water flow needed for the smooth sailing of large container ships central to global trade. Rather than simply appropriating the land for reforestation activities, as colonial businesses or conservation initiatives might have done, ForestRe (the man in the suit) would pay for locals (the *campesino* with the axe) to stop logging and do the work of tree planting. To fund reforestation activities, bonds would be purchased by those firms most at financial risk from canal closures, such as Walmart. The project is presented as a win-win-win scenario, as reforestation could yield benefits for not only Walmart but also for the environment and the local community, as the proposed reforestation will be a diverse mixture of species that local people also "find useful." The *Economist* image shows a clear-cut forest, an entire landscape reduced to stumps. According to proponents of the ForestRe initiative, what distinguishes it from other types of development, like oil palm or sugar cane monocultures, is that it is based on a broader range of services that nature provides beyond timber and other obvious commodities and, as such, will generate value while also protecting forests. The article suggests, too, that the scheme pays attention to the concerns of local people.

Taken together, the foregoing images tell a strong story of an enterprising nature: the cover image reflects the promise of orderly, efficient relations, directed by ecologies and economics; the second image of the man with the axe, meanwhile, calls attention to the processes of uneven development that run along geographical – but also classed, racialized, and gendered – hierarchies. If the two-image story were accurate and complete, enterprising nature would be a tidy set of processes that swiftly allocate resources in order to protect diverse ecosystems according to the values of northern elites. But attempts to govern are always bumpy affairs. ForestRe's 25-year bonds never materialized; the political and the economic failed to line up. Furthermore, even the ecological fact that underpinned the bond idea in the first place – the fact that reforestation activities would solve water flow issues in the canal – is now disputed.[9]

This is where the book cover image becomes inadequate, or even misrepresentative. A more apt image for this book's cover might have had the entire scene entangled by complex mangrove forest roots or fragmented into ill-fitting puzzle pieces set in a UN negotiation hall. At the very least, the man at the desk should have a sweaty brow, bags under his eyes from jetlag, and a wrinkled suit, as the creatures negotiate hard about the amount of money that goes in the bag. Enterprising nature is, I show in this book, reflective of dominant neoliberalizing processes, but nonetheless precarious, dynamic, not at all solid.

Enterprising Nature: A Dual Definition

What does it mean to call nature "enterprising"? The idea is not that nonhumans adopt business plans, take self-help courses, or hire marketing agents. The word "enterprising," as I mobilize it in this book, has two linked definitions, one as an adjective ("to be enterprising") and one as a verb, a neologism ("to enterprise").

Used as an adjective, "enterprising" describes an entity as "imaginative" and "energetic." To call someone "enterprising" is to draw attention to their creative and productive capacities, to their ability to transform something – a company, an idea, their situation – into something else, usually profitably. The adjective "enterprising" is a term typically reserved to an individual who does not depend on handouts or charity, instead relying upon their own internal smarts, ingenuity, and hard work.[10] Thus, the term "enterprising nature" refers to the way that many in the global biodiversity community represent biodiversity and nonhuman ecological bodies and processes: nonhumans *are* creative, productive, energetic contributors to life on earth.

For many proponents of this vision, the idea is not that we need to transform something uneconomic into something economic. Rather, as the head of the CBD declared, biological diversity is already the world's largest corporation; it is just not recognized as such. The task is to create the conditions wherein biological diversity can be seen for what it really is: "enterprising," with diligent work habits that can finally be recognized. This perspective runs strangely parallel, in many ways, to the views expressed by poverty experts Hernando de Soto and C.K. Prashad, and even Nobel Peace Prize–winning microcredit financier Muhammad Yunus: the world's "bottom billion" or poorest people are *naturally* enterprising. They are, as Foucault (2008) would say, "entrepreneurial selves." In this view, what people in poverty need are the conditions in which this true nature can be realized, conditions generated perhaps by extending microcredit (as Yunus has done), or by establishing the right private property relations (as de Soto proposes). In creating these conditions, the goal is to end relations of aid and charity, the "handouts" that supposedly stymie this true nature of human ingenuity, creativity, and productivity. The idea of recognizing inherent entrepreneurship is so pervasive that now, as my research shows, it is being extended toward nonhuman nature. In the era of "enterprising nature," biodiversity conservation is about creating the right conditions, the conditions under which nonhuman communities can generate the political, financial, and social capital to fend for themselves.

The multiplication of "enterprising units" is a central attribute of what Foucault, in his 1978–79 lectures at the Collège de France, terms "neoliberal governmentality" (Foucault 2003). For Foucault, neoliberal rationality and governance center on cultivating competition. This competition is not *necessarily* focused on profit-making, but on establishing quantitative differences, measuring economic magnitudes between those differences, and then using such differences to regulate and manage governing choices. Broader than a market logic, which is more tightly focused on creating the conditions for commodity production, pricing, and circulation, neoliberal rationality is focused more on rationing and optimizing "scarce resources"; it is focused on producing something like a "permanent economic tribunal for life" (2008, 247), where the most enterprising can stand up, be seen, and reap their rewards.

Enterprising nature, then, involves the creation of "enterprising units" of nonhuman life in order to set the conditions wherein *differential value* of species, nonhuman communities, and spaces can be calculated – the creation of what I describe as an ecological-economic tribunal for life (see chapter 3). This project involves rendering the qualities of ecosystems, and rich socioecological relations, into representative forms that can be compared, ranked, and ordered quantitatively. In making the world more enterprising, or competitive, the tribunal marks the "differential endowments" of diverse ecological relations in order to make biodiversity tractable for modern governance, no matter the specific breed of policy – market-oriented, command and control, or community-based.

Enterprising nature, then, is not all about the "tions" – privatization, commodification, financialization, and accumulation – at least not directly.[11] It comprises projects and initiatives that are at times about new kinds of commodities, but also about directing efficient government investments in "green infrastructure." For example, I argue that ecosystem services are better understood as a political-scientific strategy to create new interests in nature, to prevent "stupid decisions," as one advocate stated, than as an epochal transformation creating new commodities (see chapter 4). This does not mean the processes at play are benign. Nature is of course an "accumulation strategy," to paraphrase Neil Smith (2007), and has been for a long time (see Peluso 2012); it is the case that capital must find places to constantly expand. But the project of enterprising nature is better described as an attempt to manage the excesses of capitalism that are degrading life on this planet, a project driven by actors who are sometimes but very often not motivated by profit or potential of profit.[12]

Enterprising nature is more often like Tania Li's "will to improve." But unlike Li's study of development in Indonesia, where projects of improvement bracket out political-economic processes (preferring to focus on cultivating individual and community capacities to manage and adapt to their circumstances), the people and institutions at the center of this project actually take on political-economic processes, attempting to improve the functioning of market society as a whole. "We're just actually trying to make capitalism work the way it's meant to," one ecosystem service scientist and model-creator explained to me, by bringing the full ecological costs of production into state and market decision-making.[13] This is a distinctive dream for proponents of market society, one where the invisible hand efficiently directs land use and wealth distributions, a hand now operating within the (also neutral and objective) constraints of scientific calculation (see chapter 3).

Enterprising nature, then, is an attempt to bring biodiversity into governance calculations, into a kind of "permanent visibility" (Foucault 1995) for decision-making on, say, resource developments or land use decisions.[14] The circuits I study are creating the conditions that can allow biodiversity to become permanently visible not only to environmentalists but also to market relations, governance calculations, and power brokers. An enterprising nature, the trustees I study seem to be saying, will be considered *automatically* in decisions; an enterprising nature will be assessed, evaluated, and invested in neutrally and objectively, depending on its work habits and productivity. The friendly accountant pictured on the book's cover will dole out the necessary payments. Not equal payments, of course: an ecological-economic tribunal will adjudicate the enterprising nature of nature in order to guide efficient investments of scarce resources, a process that necessarily involves making new rankings and hierarchies.

"To enterprise"

Creating and multiplying these enterprising units requires incredibly dense inputs and work, including governmental interventions. This is because, as Foucault explains, competition is not a natural characteristic waiting to be unleashed, if only, for example, the state would get out of the way. Bodies are not in fact born enterprising. As such, a key aspect of the art of neoliberal government is to *create the conditions* upon which competition can be fostered. Neoliberal governmentality, then, requires not laissez-faire, but actually "permanent vigilance, activity and intervention" (Foucault 2008, 132) on the part of the state but also many other actors. This is why the accountant behind the desk on the cover should have sweat on his brow: cultivating the conditions for competitiveness to flourish, for enterprising nature, takes work; it takes interventions that are "no less dense, frequent, active and continuous than any other system" (145). It is not easy, and it involves enormous amounts of coordination. The man behind the desk on the cover of the book should be not only sweating; he should also be surrounded by experts, his position at the desk supported by institutions – academic, NGO, international. The amount of money in the bag should be underpinned by fractious governments and firms but also computer models, devised on the basis of abstract ecological-economic models.

In the second meaning of the book's title, enterprising can be understood as a verb: "to enterprise." It was long after I began using this term that I realized it was grammatically non-existent. My improper use nonetheless matches so well the processes I observed. Enterprising nature is productive; it is a set of ongoing undertakings, actions requiring ever-proliferating effort. In using "enterprising" in this way, I am also drawing from the work of theorists who call attention to the ways in which certain rationalizing knowledge systems or new representations do not somehow reveal a pre-existing world but, rather, help to bring new relations into being (Haraway 1988, 1997; Mitchell 2002, 2007). Scientists, economists, and other experts create the conditions – the subjects and objects, the regulations, laws, policies, and models – needed so that "competitive mechanisms can play a regulatory role at every moment and every point in society" (Foucault 2008, 145). "Enterprising" as a verb, then, calls attention to the *productive* work of creating a visible and economically legible biodiversity that can be seen and invested in by liberal institutions and within capitalist social relations. Global expertise actively produces "new forms of value, new kinds of equivalences, new practices of calculation," and importantly, "new relations between human agency and the nonhuman" (Mitchell 2002, 5). Enterprising nature is an *attempt* to transform, to reorganize, to produce new subjects and objects.

The work of making "enterprising nature," I suggest, traverses three interconnected processes: assembling consensus on the nature of the problem and solution across a wide variety of actors, logics, and institutions; developing methods and tools to calculate value; and building political-economic agreement that can redirect capital (state and private). The book is organized into three parts, each focusing on one of these processes. At the same time, however, these processes are densely inter-linked, functioning together "to enterprise" biodiversity conservation.

Assembling a framework

Making nature enterprising involves the production of new disciplinary apparatuses and scientific objects, ones that can bring together different logics, methodologies, and concerns to define both the commonsense problem and solution. The book's first two chapters describe the roots and history of the contemporary notion of biodiversity, pulling out some histories of the now commonsense drive to enterprise. In chapter 2, I explore the birth of global biodiversity as a scientific, political, but also deeply *economic* object in the 1980s and 1990s, an achievement involving a wide range of institutions and desires. One of the most powerful circuits involved in making nature enterprising sits at the disciplinary intersection of ecology and economics; chapter 3 explores these intersections via analysis of the Beijer Institute on Ecological Economics and its project on biodiversity in the early 1990s. At the Beijer Institute, ecologists and economists created the interdisciplinary field of "the economics of biodiversity," attempting to make biodiversity loss and conservation visible, legible, and calculable for governance processes.

Calculating value

"To enterprise" necessarily involves more than shared scientific objects like "ecosystem services" or "biodiversity"; it requires new calculative tools and models that can do the work of rendering the ecological world comparable.[15] Even if scientists and experts agree that biodiversity loss can be understood as an "externality" – a problem emerging because those who cause it directly through, say, a new soy plantation do not have to pay for those damages – the next logical step is to develop the accounting practices to internalize this externality. One needs not only a shared grammar of life but also a shared way to calculate life, to calculate the specific value of "units," of "trade-offs" that need to be made. Chapter 4 analyzes the Stanford-based computer model InVEST, which does the practical work of enterprising: calculating trade-offs between competing ecosystem services – between, say, carbon sequestration and food production, not only in the present but decades ahead. In rendering the costs and benefits of different land uses quantifiable, InVEST aims to give the state, firms, and overall "decision-makers" the ability to simultaneously govern environmentally and economically.

Chapter 5 examines a different set of calculative devices – accounting tools that attempt to suture together the ethical and biopolitical/global social reproduction interests of biodiversity conservation with financial and corporate interests – tools attempting to create an invisible hand for conservation. The chapter takes a close look at two calculative devices, the Integrated Biodiversity Assessment Tool and the Corporate Ecosystem Service Review. They aim to attune firms to the *risks* faced from changing ecologies, with the overarching goal to incentivize corporate investment in risk mitigation, in ecological investments (or so the theory goes), reproducing the conditions of capitalist production (O'Connor 1998). If the man at the desk on the cover of the book is the CEO of, say, Mondi, a South African pulp and paper company, then tools like the Corporate Ecosystem Service Review do the accounting work that brings various nonhuman bodies into his vision and hopefully into the line-up for payment.

Redirecting capital

Making the "big big money" flow from biodiversity – as one investment banker described their interest – requires more than frameworks or new models that show nature's true enterprising nature. Facts and numbers do not perform on their own; advocates of enterprising nature need to produce new institutional arrangements that can bring state or profit-seeking capital in line. This is my central interest in chapter 6, where I take readers into a series of "Biodiversity and Ecosystem Finance" conferences where participants grappled with how to transform biodiversity conservation into "a valuable business proposition" in order to channel capital into the good work of conservation. Chapter 7, the final empirical chapter, focuses on intergovernmental debates over market mechanisms for conservation under the auspices of the UN Convention on Biological Diversity (CBD) and on one attempt to make a globally tradable, UN-approved "asset in biodiversity." Although a tradable asset in global biodiversity appears as the logical progression from the early ecology-economics synthesis of the 1990s – as perhaps the pinnacle of the ongoing work to create "biodiversity" as an object that can be seen by market and state institutions – the CBD initiative turns out to be a fraught and controversial endeavor, failing to receive intergovernmental assent.

The Friction-Filled Terrain of the Neoliberal, the Universal, the Global, and the Enterprising

Enterprising nature is part of the broad story of socioecological re-regulation taking place since the 1970s, what critical scholars and activists call neoliberalism.[16] It is a part of a global and universal dream to "improve" or, in the words of several of my interviewees, to "stop making bad decisions." It is also linked to neoliberal policies and practices that function to expand the conditions for elite profit, a point well-made by David Harvey. Interventions by academics like Kathleen McAfee, Neil Smith, Sian Sullivan, Bram Büscher, and Morgan Robertson emphasize that biodiversity conservation is becoming sutured to accumulation processes, "accumulation by conservation" (Büscher and Fletcher 2015, 202), intensifying "the commodification of life itself" (Sullivan 2013, 210).[17] International activists also make use of such framing. A wide-ranging group of organizations including Friends of the Earth International, Ecologistas en Acción and Carbon Trade Watch recently produced a video titled "Stop the Takeover of Nature by Financial Markets," a video that interprets what they call the "financialization of nature" such as the carbon or biodiversity markets as driven by the incessant, restless logic of capital, by capital's need to find new spaces and new bodies to fuel accumulation.[18]

I share these concerns.[19] Yet, enterprising nature is different than the financialization of loans, mortgages, or even food; it involves a different and diverse set of actors, institutions, and driving rationalities. Critical scholars and activists studying these processes (myself included) can be too quick to characterize them as having a "coherent ideological and institutional formation with necessary outcomes," a criticism Wendy Larner (2007, 220) makes of the neoliberal environments literature. She worries that there is a tendency to represent neoliberalism as monolithic and unstoppable. Similarly, the "accumulation by conservation" lens can be teleological – as in, *of course* biodiversity conservation is a target for capital's never-ending search to accumulate, a process driven by shadowy, unspecified capitalist elites with singular interests. (Illustratively, in the video I mention above, the protagonist is the mustached businessman from the Monopoly game, known as "Rich Uncle Pennybags" or "Mr. Monopoly.")

Any analysis of enterprising nature, then, faces a long-standing tension in critical theory: how to research, talk about, and represent the actually existing structuring forces of global capital, of the global and the universal, in a "way that avoids lending them a logic, energy and coherence" (Mitchell 2002, 14), that avoids telling a story of the unfolding potential of "the global forces of modernity, of science and technology, and of the expansion of capitalism" (2002, 13–14). In one view, held by some trustees of enterprising nature, the ascendancy of the "to make live, one must make economic" maxim in global biodiversity conservation is the natural progression of the free market, part of the crafting of a perfect capitalist system that can account for *all* environmental ills and allocate resources, including ecological functions, efficiently. We need to avoid a similarly teleological critique. When we let the "global appear homogenous," Anna Tsing (2005) persuasively argues, "we open the door to its predictability and evolutionary status as the latest stage in macronarratives" (2005, 58).

So again, how can we account for global forces, for deep uneven clusters of power and knowledge, for pervasive and structuring "universal dreams and schemes" (Tsing 2005, 1) without letting the global dominate the local, without letting capital, the colonial, and the Global North yet again define the terrain?

One crucial way researchers produce a more critical, complex account is by examining what happens when the "universal or the global" hits the ground, and by outlining the hybrid forms of neoliberal governance that emerge as a result.[20] Li calls this the moment when expert discourse or the plan, what she calls the "practice of government," comes into contact with the "practice of politics" (12), the "witches' brew of processes, practices and struggles that exceed their scope" (28).

Complementing this attention to "hybrid neoliberalisms" on the ground, I draw out how the global, universal, and neoliberal discourse of enterprising nature is itself a "witches' brew." Like local or national formations of rule, global governance is not homogenous, coherent, or straightforward, but rather "charged and enacted in the sticky materiality of practical encounters" (Tsing 2005, 1). The global dream of enterprising nature is not abstract or inexorable; it is produced by specific people, institutions, epistemological frameworks, computer models, databases, laws, and policies. As Geraldine Pratt (2004) explains, tracing the production of discourses as "situated practices in particular places" (20) is one way that we can make contradictions and tensions come to light, and thus also is a way to create conditions for political possibility, for political agency. And indeed, the "practical encounters" I trace are not at all smooth, but rather loaded with frictions, unknowns, leaps of faith, and disagreements.

To be clear, my intervention is not simply to call for more attention to "non-capitalist" spaces or logics (e.g. Gibson-Graham 1996). Neoliberal capitalism is a dominant mode of production on planet earth; enterprising nature cannot be understood separately from its hegemony.[21]

Yet, not all that happens under the banner of enterprising nature can be understood as a response to the unfolding logic of capital, or a singular pursuit of profit. Rather, there is a much more "unwieldy and contradictory political assemblage" (Larner 2007, 220) that must be wrestled into coherence, an arduous, even sweaty process. Actors have to try to align multiple desires, logics, rationalities, and interests.[22] Again and again, they have to try to reconcile the fissured desires of ecologists (including scientific curiosity, research funding, and ethical drives), of conservationists struggling within changing institutional mandates, of bureaucrats who are trying to make inroads into increasingly closed national budgets, and of green entrepreneurs who sometimes want to make a quick buck and other times are in it for the long haul. To govern, to nail down the right relation between humans, or between humans and diverse others, is not to "seek one dogmatic goal" but rather to achieve, as per Foucault, "a whole series of specific finalities" (Li 2007, 9). Enterprising nature is well steeped in capitalist social relations, no doubt, but these social relations are forged from multiple logics that are not always oriented around the singular pursuit of accumulation. Many of those doing the work of "enterprising," such as Pavan Sukhdev and Stanford ecologist Gretchen Daily, while certainly not anti-capitalists by any stretch of imagination, do not hold the same interests as the CEO of Shell. Practical encounters among ecologists, economists, financiers, and bureaucrats working in disparate institutions with not entirely aligned goals result in cobbled together coherences loaded with what Tsing terms "frictions."

"In the middle of things"

The primary goal of this book is to trace the apparatus that gives life to enterprising nature, to locate the roots and perpetuating practices of this increasingly commonsense, increasingly hegemonic approach. Following Foucault (1977, 194), I understand the term "apparatus" to mean the relations both material and semiotic among "thoroughly heterogenous" people, institutions, capital flows, ideas, regulations, science, valuation methodologies, computer models, and databases – relations that together produce a particular idea of a problem.[23] To trace the apparatus-in-the-making of enterprising nature, I travel among specific moments where biodiversity is tethered to economic logics, to show the specific people, institutions, and epistemologies involved. As geographer Stephen Legg points out, while apparatuses have coherences, they are not closed. Rather, as Legg writes, the very "multiplicity" of apparatuses, by which he means the diverse set of rationalities, actors, and practices that give them life, "necessarily opens spaces of misunderstanding, resistance and flight" (2011, 131–2).

Enterprising nature does not take shape in a single location. As such, my research took place in multiple sites and relied on multiple methods as I sought to trace what I call, drawing on Roy (2010), transnational "circuits" of power and knowledge. These circuits are upheld by many actors – scientists, policy wonks, economists, entrepreneurs, and financial managers – operating from diverse institutional bases, including universities, government, business, the non-profit sector, and intergovernmental organizations. Enterprising nature is an interdisciplinary, highly collaborative, transnational project that brings together both likely and unlikely allies. I opt for the term "circuits" (Roy 2010) over the more commonly used "networks" because "circuits" suggests that the routes travelled are well-worn and regularized. Enterprising nature is a world where the same institutions and people appear over and over again. The circuits I trace are international, but they are resolutely not *global*: these circuits are busy pathways among a handful of academic institutions and international organizations predominantly in the United States and Europe.

In crossing between various sites, I also crossed between a diverse set of rationalities, actors, and practices that did not always line up. I was constantly surprised: a conservationist I thought would be aligned with a new initiative to economically value biodiversity would go on to question its efficacy. An ecologist whose published work appeared to be producing new methodologies for the commodification of nature would disagree vociferously with my interpretation in an interview, expressing deep reservations about the entire project of enterprising nature.

My approach to studying these circuits involved living in them as much as possible, beginning, as Anna Tsing describes, "in the middle of things" (2005, 1). I travelled to places like Geneva, Cambridge, Washington, New York, and Palo Alto. I interviewed a wide range of actors, especially scientists and experts, in their offices, in coffee shops, via Skype, while strolling through gardens in Nairobi, and, in one case, in a peanut bar.[24] I had in-depth interviews with many of the most ardent and prominent supporters of an economic approach to biodiversity, and I sought to find out their rationales and motivations; I dug into new economic-ecological models that aim to tell us the return on conservation investments decades into the future; I observed environmental market-promoters in their own "habitat" at conferences.

My research also draws on a 10-year engagement with international civil society groups working on global biodiversity policies. As I discuss in the preface, this research emerged out of pre-existing collaborations with a range of environmental organizations and social movements.[25] I conducted my research as a participant in the circuits of knowledge and power that give shape to the global biodiversity apparatus. For me, then, beginning "in the middle of things" also meant organizing and facilitating strategy meetings for the CBD Alliance and developing joint policy papers across organizations as different as the World Wildlife Fund and the International Collective in Support of Fishworkers.[26] Being in the middle of things involved listening to government delegates at negotiations make arguments and counter-arguments, often late into the night; it even meant crafting suggested alternative decision text and circulating these suggestions to friendly government delegates.

As in most field research, my position in "the middle of things" was not entirely comfortable.[27] In the course of my research I often found myself, as Roy (2010) nicely describes, in zones of awkward engagement, engagements where I was constantly without a clear identity. Was I a researcher? An advocate? Was I a programmer or critic? (This is a distinction Tania Li (2007) makes between those who design and implement projects and critics who stand at arm's length – a distance she deems necessary for critique.) For me these positions were not mutually exclusive, and through them I came to see the complexity, challenge, and uncertainty within the global, the universal, and the neoliberal of enterprising nature, a project that is coherent but very far from closed.

Four Tensions within Enterprising Nature

Productive engagement with enterprising nature means working within a series of persistent tensions in contemporary conservation. I introduce these tensions as four complex and often contradictory realms of thought that the book will analyze in depth. Any participant in contemporary biodiversity politics will find herself navigating these tensions; there is no pure, uncomplicated position from which to work.

In what ways does enterprising nature both politicize and de-politicize conservation?

Enterprising nature is part of a major rethink taking place in mainstream conservation: a "post-natural" turn within which conservationists have turned away from trying to protect "pristine" nature as a realm outside human life and instead focus on protecting the parts of nature that best contribute to human well-being.[28] Ecosystem services – a study of the benefits to humans of particular ecosystem functions – is one example of the new approach. In this post-natural turn, conservationists ask very political questions: For whom do we conserve? How can we achieve more just conservation? In this sense, the new conservation is politicizing – focused on where and to whom the benefits of particular natures flow.

At the same time, however, enterprising nature attempts to create a new universal in conservation, a neutral, objective, apolitical approach that can determine the value of particular socioecologies. The idea is to situate conservation within the supposedly *undeniable* grammar of quantitative value; initiatives within the enterprising nature project seek to solve complex problems of socioecological justice by transforming them into questions of accounting, with accounting systems designed by an elite group of Northern experts. Thus, while the post-natural turn appears to promise conservation within a more human-centric, poverty-alleviating framework, enterprising nature reveals the persistent tendency for global environmental initiatives to be defined in "the superior economic and institutional power of Northern parties and Northern-based NGOs" (McAfee 1999, 140). In this sense, enterprising nature is intensely *depoliticizing*, turning away from responsibility and justice through the development of financial mechanisms (see chapters 5, 7).

Is the ranking of socioecologies emancipatory or exclusionary? Is it even possible?

Conservation has always been selective in investments of time and resources; most well-known is its longstanding preference for charismatic mega-fauna. An enterprising nature could potentially broaden social concern for a wider array of nonhuman species. For example, an ecosystem service frame might turn more attention to the work of microbes and other decomposers that are crucial in soil health. At the same time, however, ecologists and other advocates of enterprising nature seek to be strategic. If resources and political will for conservation are scarce, investments must be efficient and selective; it is not efficient, rational, or even possible to value all species equally, especially in times of austerity. Thus, in seeking to protect forms of nonhuman life, enterprising nature works to create hierarchies, to create and calculate differential values of socioecological relations. Enterprising nature could produce something like surplus non-human populations: those deemed not necessary, or redundant, or at the least, less worthy of investment. The result is a kind of triage, an instru-mental and economic approach to deciding which species matter and which can be more marginal to human concern. Thus, in seeking to broaden the focus of conservation, enterprising nature may also create new exclusions.

Yet we must also ask to what extent such rankings are even possible. Every expert I spoke with over the course of my research emphasized how difficult this form of quantification remains; both the ecology and the economics remain fraught with uncertainty and unresolved complexities. How do individual species contribute to ecosystem functioning, and then, further, to human well-being in the form of services? As one academic ecologist noted in an interview, "The rela-tionship *between* biodiversity and ecosystem services is an area we need a lot of research in ... We don't know what diversity does." Biodiversity is incredibly dynamic, animate, complex, and therefore immensely challenging to represent and measure. As such, contemporary understandings of biodiversity largely lack the systems of representation to value and rank different parts of the nonhuman world, to transform the complexity of life into discrete ecosystem services.

What are the risks and opportunities of endless ecological simplifications?

Ecology is a field known for increasing scientific understanding of the immense complexity of environmental processes. For the past several decades, however, ecologists and conservationists have tried to bring complexity and uncertainty into the sights of liberal institutions. In return, these institutions demand further and further simplification; ecologists are asked again and again to reduce the complexities in their models, to make their findings more straightforward and legible to governance. When it comes to bringing biodiversity into models of economic valuation that decision-makers might recognize, ecologists face tough choices: "You're basically going to accept imperfect proxies," one academic told me, some "pretty ugly trade-offs." Models used rest on layer upon layer of abstraction (see chapters 4, 5).

Thus, while enterprising nature wants to bring ecological complexity into governance (state, finance, corporate), the project also bears similarities to the much-maligned models of maximum sustainable yield (MSY): both aim to render future ecological-economic futures known. There is a risk, then, that enterprising nature is a "complexity blinder," as Richard Norgaard (2010, 1219) calls the ecology of ecosystem services. Enterprising nature continues to seek the ideal model, the perfectly simplified synthesis of ecological and economic knowledge that will offer decision-makers definitive answers about the "best" course of action; that model, of course, rests on an ever-retreating horizon. Ecologists remain caught in the ongoing tension of trying to share their understandings of complex ecosystems in a way that will be taken up in liberal governance. Yet this tension is regularly addressed not by questioning, say, the nature of liberal institutions demanding these simplifications, or by assessing the fundamental limitations of the project, but rather by calling for more interdisciplinary knowledge, more investments in expertise, or perhaps a new model that can bring the unruly unknowns of socioecologies into order.

What is the "human place in nature"?

"No one cares about nature," says Pavan Sukhdev. For Sukhdev and his enterprising nature allies, the only viable conservation approach requires a focus on the utilitarian, the use values of nature, ideally transformed into quantitative and perhaps monetary form. As such, enterprising nature is not only about creating the conditions for biodiversity to be automatically and efficiently considered in governance, but also a particular way of conceptualizing the "human place in nature" (to borrow a phrase from the subtitle of William Cronon's 1995 book). This place in nature, ironically – at least for environmentalists – is one that aims to make nonhumans more "passive" and "tractable" for calculation (as Lohmann 2009, 503 puts it), while at the same time hoping that this increased instrumentalism will result in *less* human domination and devastation of nature; that it will rectify a "deep denial of human dependence on nature" (Plumwood 2002, 71).

Attempts to create new "interests" in nature via ecological-economic modeling may also have the effect of creating further abstractions, further distances, perhaps even flattening and deadening nature (Büscher et al. 2012, Sullivan 2013, Igoe 2012). As a broader strategy, then, this approach seems oppositional to activist and critical academic thinking on the role that human domination and dominion have played in legitimizing environmental and social devastation and injustices (e.g. Horkheimer and Adorno 1944, Plumwood 1993).

Yet, there *is* a crucial need to shorten the distance between people, especially Northerners, and the close-but-also-very-distant socioecological sites that sustain them. The global political economy has for too long relied upon sacrifice zones, on sacrificed bodies, on sacrificed futures: places and bodies and relationships which bear the brunt of socioecological changes, over and over again (Klein 2015). A huge shift in political economic relations is needed to break these patterns, patterns that are a part of explaining the monoculturing of the planet. The question remains, however, is enterprising nature up to the task?

Notes

1 Statistics from Butchart et al. (2010), Hoffman et al. (2010), Dirzo et al. (2014), FAO (2010), Dulvy et al. (2014).
2 This mantra – "To make live, one must make economic" – is inspired by Michel Foucault's description of biopower, a form of modern power that operates to "make live and let die" (see Foucault 1990 [1978], 2003). This is a modality of power focused on regulating, administering, and managing life to secure and optimize the health of the population as a whole (the "whole" of concern for Foucault, especially as expressed in *Society Must be Defended*, is that of the nation state). Historically, Foucault demonstrates the emergence of biopolitical forms of power through the emergence of new knowledges about the population: birth and death rates, illnesses, economic indicators – especially measured in statistical terms, and also linked to interventions aiming to

adjust and control "macro" processes (i.e. birth control, public hygiene, insurance). This includes, as the "last domain" of biopolitics, "relations between the human race ... and their environment, the milieu in which they live" (Foucault 2003, 245). As I outline in chapters 3–5, enterprising nature is very much focused on securing the health of "the whole" through the economical and ecological management of nonhuman bodies and socioecological processes via new knowledges, models, and accounting devices. However, the health of the "the whole" is not limited to the nation; the target can be humanity (chapters 2, 3), a region or national polity (chapter 4), or the firm (chapter 5). Finally, a biopolitical approach does not at all mean that all lives – human or nonhuman – are equally valued or fostered. Some lives are more or less expendable, and killing is permissible when it results in "the elimination of the biological threat to and the improvement of the species or race" (Foucault 2003, 256), a point Foucault demonstrates with his critical discussion of state racism in *Society Must be Defended*. In enterprising nature, too, some lives are designed as more or less expendable, some spaces are more or less investable, this time depending on the results of an ecological-economic tribunal.

3 While cumbersome, the term socioecological signifies the inextricability of the social and the ecological, a refusal to separate nature and society, or the human and nonhuman. Similar neologisms include "naturecultures" (Haraway 2003), "social nature" (Castree and Braun 2001), and "humanity-in-nature/nature-in-humanity" (Moore 2015).

4 A "god trick" is Haraway's (1988) critique of objectivity in science, in that it presents itself as synoptic and impartial, a "view from nowhere." The problem with the god trick, for Haraway, is that knowledge claims that come from nowhere cannot be held responsible for their effects. Here I am suggesting that the god trick of scientific objectivity is being combined with another kind of science from above, that of neoclassical economics, which also makes claims to neutrality based in the operation of price signals. I describe this point in chapter 3.

5 Bram Büscher, Sian Sullivan, and their co-authors (2012) argue that an effect of the marketization and commodification of biodiversity conservation, what they term neoliberal conservation, is that "nonhuman natures tend to become flattened and deadened into abstract and conveniently incommunicative and inanimate objects, primed for commodity capture in service to the creation of capitalist value" (23). See also Sullivan (2013).

6 This is a central argument of Frankfurt School theorists Max Horkheimer and Theodor Adorno (1944), as well as feminist scholars like Donna Haraway and Val Plumwood.

7 By liberalism I am referring to "branches of European political philosophy that include the narration of political emancipation through citizenship in the state, the promise of economic freedom in the development of wage labor and exchange markets, and the conferring of civilization to people educated in aesthetic and national culture – in each case unifying particularity, difference, or locality through universal concepts of reason and community" (Lowe 2014, 3–4). See also Domenico Losurdo's (2014) excellent book *Liberalism: A Counter-History*.

8 The example of John Locke is helpful here. Locke himself was at the epicenter of emerging liberal ideals promoting economic and political individual freedoms, but was an active investor in and beneficiary of the slave trade. In the American Revolution, those who supported the revolution as an act of liberty were not necessarily abolitionists, a situation Domenico Losurdo argues can be explained by the material interests of patriots in maintaining slavery.

9 A recent paper in the *Proceedings of the National Academy of Sciences* raises questions about the level and extent to which reforestation increases water supply to the canal (Simonit and Perrings 2013).

10 I am indebted to the prescient work of Kathleen McAfee on this point. While her seminal 1999 essay, "Selling Nature to Save It," was focused specifically on bioprospecting as an approach to conserve biodiversity, McAfee writes that such an approach *"offers to nature the opportunity to earn its own right to survive in a world market economy"* (134; her emphasis). I first read McAfee's essay after I had participated in my first CBD meeting in 2002, just as I started my Master's degree. I remember feeling that she was onto something crucial, especially in noting a trend towards environmentalism coming to "service of the worldwide expansion of capitalism" (134). In many ways my book is an exploration of her arguments in that 1999 article, the post-bioprospecting attempts to make conservation pay, although my research approach and the 15-year period between her article and this book lead me to different conclusions.

11 Robert Fletcher argues a similar point in a (2010) article, noting the need to understand what he calls neoliberal environmentality not as "merely a capitalist economic process," but rather as a more "general strategy for governing human action in a variety of realms" (171). Drawing from Foucault extensively, Fletcher's article helpfully articulates the varieties of power operating in conservation.

12 James O'Connor (1998) terms these excesses of capitalist social relations the "second contradiction," by which he is referring to the tendency for the biophysical conditions for capitalism to be degraded by capitalism itself.

13 Pavan Sukhdev, the head of The Economics of Ecosystems and Biodiversity (TEEB) project to study the economic impact of the global loss of biodiversity, at times sounds revolutionary, writing that "the root causes of biodiversity loss lie in the *nature* of the human relationship with nature, and in our dominant economic model," going on to say that our current economy "promotes and rewards *more* versus *better* consumption, *private* versus *public* wealth creation, *human-made* capital versus *natural* capital" (TEEB 2010b, xviii, his emphasis). Yet, despite this widespread criticism of contemporary Western culture and economies, in the next breath Sukhdev states that the main problem is market-failure: "there are no 'markets' for the largely public goods and services that flow from ecosystems and biodiversity" (TEEB 2010b, xxi).

14 The terminology "permanent visibility" comes from Foucault's (1995) discussion of the Panopticon in *Discipline and Punish*. The Panopticon is a building designed by Jeremy Bentham in the late-eighteenth century that allows

an observer to observe all inmates of an institution without them being able to tell whether or not they are being watched. The brilliance of the Panopticon, for Bentham, is the way that it is a form of power, or a technique of power, that achieves its ends efficiently, in that it does not rely upon power added from the outside, "like a rigid, heavy constraint." As a form of power, panopticism is more subtle, built into the mode of visibility itself, operating almost automatically in an attempt to discipline subjects.

15   Calculation is the process of establishing "distinctions between things or states of the world," as well as "imagining and estimating courses of action associated with those things or with those states as well as their consequences" (Callon and Muniesa 2005, 1231, drawing from Latour 1987). Calculation is not a "universally homogeneous attribute of humankind, nor an anthropological fiction" (Muniesa et al. 2007, 5), but rather a "concrete result of social and technical arrangements," arrangements that can be described as "calculative devices." Calculative devices make it possible to compare and contrast different courses of action, economically, and although technical, they are always political. Calculative devices "do things," as performativity of economics scholars Fabian Muniesa and his co-authors state (2007, 2); they have effects in the world by translating rich qualitative relations into hard, quantitative numbers.

16   There are many definitions of neoliberalism, stemming from the many books devoted to the concept (e.g. Harvey 2005, Peck 2010, Brown 2015, Mirowski 2013), not to mention all the articles. Geoff Mann's (2013) definition is one I regularly turn to: "Neoliberalism is the ongoing effort, in an inevitably uneven global political economy, to construct a regulatory regime in which the market is the principal means of governance and the movement of capital and goods is determined as much as possible by firms' short-term returns" (148). Neoliberalism, then, refers to political-economic reassertion of elite power and profit, beginning around the early 1980s ("creating the conditions for short-term returns"). It is also about producing subjects and relations of rule that are market-like or economic in form; as Wendy Brown insists, neoliberal ideologies and practices are a political rationality that reaches "from the soul of the citizen-subject to education policy to practices of empire" (Brown 2005, 39). By its very name, neoliberalism suggests a continuation of older logics and processes. Its most sacred principles include private property, individual freedom, a state whose main role is to protect these property rights and freedoms, and a laissez-faire approach to environmental regulation in order to facilitate economic development. These principles stem in particular from classical liberalism, a Western political ideology that is classical because it pre-dates the modern age and liberal because it holds that "the golden road to collective wealth" is through individual freedom and a society unconstrained by the state (Mann 2013, 142).

17   There is a wonderful and growing body of literature here. McAfee (1999) tracks the rise of green developmentalism, a market solution that aims to literally "sell nature to save it." Smith (2007) discusses the emergence of carbon trading schemes and other ecosystem service markets as signaling

not only capital's pursuit of new ways to accumulate but a means of subsuming biological processes to capital (Smith draws from Boyd et al. (2001) to interpret these developments as indicative of a shift from formal to real subsumption of nature). Morgan Robertson (2012) extends Smith, arguing that the commodification of ecosystem services signals a transformation comparable to that of individual human labors becoming social labor under capitalism, heralding new forms of accumulative processes which generate profits not only off of nature's goods but also its services. Others working in what is called "neoliberal conservation" also interpret the rise of market rhetoric among practitioners as reflective of a broad repurposing of conservation around the logic of capital, which as Büscher et al. (2012) write, "shifts the focus from how nature is used in and through the expansion of capitalism, to how nature is *conserved* in and through the expansion of capitalism" (Büscher et al. 2012, 4, emphasis added). Much of the work in this area links to David Harvey's theorization of spatial fixes but in an environmental register, explaining market solutions as a "fix" to capital's "constant need ... to expand its reach into new spheres of accumulation" (Arsel and Büscher 2012, 57; see also Brockington and Duffy 2010, Büscher et al. 2014). This line of reasoning emphasizes the role that new environmental markets – in this case, structured around the management of biodiversity conservation – play in transforming nature into an "ecological" fix for capitalist crises of accumulation (Castree 2008a). The proliferation of market-driven conservation strategies and tradable environmental commodities is understood (again drawing on Harvey) as a new but "similar and spectacularly productive" (Sullivan 2013, 210) wave of accumulation by dispossession. On this and neoliberal conservation in general, see also Büscher (2009), Arsel and Büscher (2012), Brockington and Duffy (2010), Igoe et al. (2010), MacDonald (2010), Fletcher et al. (2014), Büscher (2009, 2014), Igoe and Brockington (2007), McAfee (1999), MacDonald (2010), Büscher and Fletcher (2015).

18 View the video at https://vimeo.com/43398910 (last accessed February 11, 2016).

19 For example, I've co-written briefing notes making similar points for Convention on Biodiversity negotiations (i.e. CBD Alliance 2010) as well as academic articles with the title "Life Is Not for Sale" (Collard and Dempsey 2013).

20 See for example McAfee and Shapiro (2010), Matulis (2013), Fletcher and Breitling (2012), Shapiro-Garza (2013).

21 I'm grateful for discussions with Rosemary Collard and Juanita Sundberg about this point (see Collard et al. 2015).

22 Morgan Robertson's scholarship on wetland banking is exemplary in this regard, demonstrating that such translation is needed to produce a wetland banking credit that moves between the logics and practices of the law, science, and the market (see Robertson 2006, 2007).

23 Foucault (1977) defines an apparatus as a "thoroughly heterogeneous ensemble consisting of discourses, institutions, architectural forms, regulatory decisions, laws, administrative measures, scientific statements,

philosophical, moral and philanthropic propositions – in short, the said as much as the unsaid ... The apparatus itself is the system of relations that can be established between these elements" (194).

24 I conducted interviews with individuals associated with the following organizations and institutions:

- Non-governmental organizations: Forest Trends, World Wildlife Fund (WWF), Royal Society for the Protection of Birds (RSPB), Packard Foundation, World Resources Institute, Birdlife International.
- Intergovernmental organizations and government: International Union for the Conservation of Nature, United Nations Environment Programme Finance Initiative, The Netherlands Ministry of Housing, Spatial Planning and the Environment.
- Financial – private sector: World Business Council for Sustainable Development, International Finance Corporation, BC Hydro, EBG Capital.
- Academic: University of British Columbia, Stanford University, Gund Institute, Duke University, University of California–Berkeley.

Some interviewees agreed to a full use of their names within my book, whereas others requested confidentiality, or else for consultation on direct use in publications. I decided largely to make the interviewees confidential in the text, except where the person agreed and/or where the person's identity is obvious.

25 My work with the CBD Alliance involved attending 13 international negotiations of the CBD; producing informational and advocacy documents, including the widely read civil society dossier *ECO*; organizing workshops and strategy sessions; coordinating increased participation of Southern and Indigenous civil society representatives; and meeting with the Secretariat of the CBD to develop strategies to improve civil society participation in the Convention.

26 In preparation for the 9th (2008) and 10th (2010) Conference of the Parties to the CBD, I produced joint policy papers with over 40 NGOs.

27 For example, at times my political commitments came into conflict with my research. At one point the daily lobby document I co-edited, the *ECO*, published an article critical of the Green Development Mechanism (GDM) (discussed in chapter 7) at a negotiation of the CBD in Nairobi, Kenya. My association with the publication meant that I was unable to interview a key proponent of the policy mechanism.

28 See for example Karieva et al. (2012), debates in *Animal Conservation* 17(6), academic analysis of this post-natural turn be found in Robbins and Moore (2013), Robbins (2014), Collard et al. (2014).

2

The Problem and Promise of Biodiversity Loss

Biodiversity is often defined in straightforward technical terms, referring to genetic, species, and ecosystemic diversity on earth. This chapter shows how very un-straightforward biodiversity is as it becomes institutionalized in US federal politics and within a major United Nations convention. With an eye on the tensions between North and South, I trace the early ascendancy of biodiversity, an ascendancy forged through articulations between global biopolitical concerns, state-capital accumulative logics, and national security interests. The chapter begins at Stanford, with the work of ecologists Paul and Anne Ehrlich, who identify biodiversity as a global concern. The chapter ends, though, in India, with the writings of scientist, feminist, and activist Vandana Shiva, who contests the imposition of Western science and policy in the Global South.

Rivets Are Popping All Over the Place: Extinction as Global Biopolitical Concern

In 1981, ecologists Paul and Anne Ehrlich published *Extinction*, the book that identified species diversity as critical to global ecosystem functioning and, in turn, the survival of humans. *Extinction* begins with a powerful metaphor. Imagine, suggest the Ehrlichs, that you are walking toward an airplane on which you plan to fly. "You notice a man on a ladder busily prying rivets out of its wing," they write. "Somewhat concerned, you saunter over to the rivet popper and ask him just what the hell he's doing." The man says he works for the airline "Growthmania

Enterprising Nature: Economics, Markets, and Finance in Global Biodiversity Politics, First Edition. Jessica Dempsey.

Intercontinental" and "the airline has discovered that it can sell these rivets for two dollars apiece." He goes on: "I'm certain the manufacturer made this plane much stronger than it needs to be, so no harm's done. Besides, I've taken lots of rivets from this wing and it hasn't fallen off yet" (Ehrlich and Ehrlich 1981, xi).

The Ehrlichs' argument is that the loss of rivets is analogous with "the extermination of species and populations of non-human organisms" (xii); a certain number of each of these entities is essential, but which ones and in what amounts is unknown. Species extinctions here or there might not be so problematic, the Ehrlichs admit, but the cumulative effects could be catastrophic at the scale of the biosphere. Thus, "by deliberately or unknowingly forcing species to extinction, *Homo sapiens* is attacking *itself*; it is certainly endangering society and possibly even threatening our own species with extermination" (6). Because we cannot know which rivet is critical – which rivet is "the one" – we must embark from the premise that each life form or species has a specific but unknown role in maintaining ecosystems on earth. The result is an irresolvable uncertainty about how individual species relate to the function of ecosystems more broadly; we cannot distinguish species that are necessary from those that are ostensibly surplus, and, as such, we need to practice precautionary conservation, ending extinction. Securing the health of the human body politic, then, means conservation of all species.

The Ehrlichs never use the term "biological diversity," but their work is both critical to and representative of the ascendancy of the idea. That human life on earth depends on a diverse assemblage of nonhumans is also the key message of two other key texts at this time: Norman Myers's *The Sinking Ark: A New Look at the Problem of Disappearing Species* (written in 1979) and a 1980 International Union for the Conservation of Nature and Natural Resources (IUCN) report titled, boldly, *World Conservation Strategy*.[1] These texts also frame extinction as a risk that threatens "life support systems ... on which human survival and development depend" (IUCN 1980, vi). The argumentation is resolutely utility focused, centered on resource losses such as the loss of soils from deforestation, the reduction of fuel wood, the loss of fisheries capacity, and reductions in genetic diversity. The argument that diversity underpins ecosystem functioning and thus "human well-being" comes from a line of ecological thinkers in the second half of the twentieth century. Aldo Leopold, Charles Elton, Rachel Carson, and David Ehrenfeld all viewed species diversity as a crucial foundation for human and planetary health. "To keep every cog and wheel," Leopold famously wrote, "is the first precaution of intelligent tinkering" ([1949] 1966, 190).

Together, in the early 1980s, *Extinction*, *The Sinking Ark*, and the *World Conservation Strategy* position the loss of species diversity as a global biopolitical concern: "global" because the risk extends beyond the boundaries of any one nation state (the unit of Foucault's analytics) to *Homo sapiens* as a whole; "biopolitical" because extinctions are explained as a risk to the health of humanity, as undermining conditions that sustain human life.[2] In short, these books frame extinction as – to use Melinda Cooper's terminology – a "crisis of social reproduction," one that threatens "no less than the continuing reproduction of the earth's biosphere and hence the future of life on earth" (2008, 16).

Paul and Anne Ehrlich work at the Department of Biological Sciences at Stanford University, a department home to several influential ecologists, including, alongside the Ehrlichs, Hal Mooney and Gretchen Daily; all of these scholars figure highly in the biodiversity politics I track in this book and they reappear in subsequent chapters. The Ehrlichs, thus, are embedded in a powerful network of scientific thinkers within which the concept of biodiversity develops and spreads. I linger on their arguments because through their work I see both the promises and perils of Western ecological thinking and practice.

Extinction's arguments about ecological dynamics and diversity demonstrate – at least in part, I argue – what feminist philosopher Lorraine Code calls the "promise of ecological thinking." Code argues that ecological thinking is disruptive to the hubris of the Enlightenment because it is suspicious of "universals and closure," rejects the reductionism of Western science, and, more broadly, opposes the "single-minded drive towards mastery by any means" (2006, 8–13). Code argues that the promise of scientific ecological thinking is on full display in Rachel Carson's work, beginning in the 1960s with her *Silent Spring*. I suggest it is also present in the Ehrlichs' rivet hypothesis, where the ecological dynamics and relationships are positioned as "known unknowns." The rivet hypothesis suggests that the whole cannot be simply apprehended from its parts; it rests on enormous uncertainty, dynamism, and surprising relationships among organisms and between organisms and their environments. In their emphasis on "known unknowns," the Ehrlichs represent the planet as ontologically dynamic and uncertain. As such, *Extinction* offers some disruption of the rationalist worldview that nature can be objectively seen, known, and controlled.

In *Extinction* the Ehrlichs also take aim at economics, denouncing mainstream economists for lacking any sense of limits. "Decision makers," the Ehrlichs write, "do not recognize the urgency of getting on with it" – that is, of facing up to the problem of species extinction – "in part because the economists who advise them are utterly ignorant" (1981, 245). To the Ehrlichs, economists are ignorant because they believe that everything, including living things and systems, are substitutable and that material growth can be limitless on a limited planet. The Ehrlichs reject the "just set aside a preserve" approach as wholly inadequate. Drawing from Herman Daly's *Steady-State Economics* (1977), they call for dramatic reductions – reversals, in fact – in rates of both population and economic growth. They blame "rich nations" (and their econo-mists) and call for "a retreat from overdevelopment" (246); they partic-ularly criticize international development institutions such as the World Bank for supporting expansion of monoculture agriculture and cattle ranching that cause tropical deforestation. Redistribution is an "essential ingredient for achieving a sustainable society"(247), the Ehrlichs explain, as is rapid dematerialization of the economy. All of this, they say, is "clearly feasible in a democratic, essentially capitalist system" (246).

Read generously, then, the Ehrlichs' *Extinction* holds some of the promises of ecological thinking. It is a book that, along with others written at the time, at least to some extent, unsettles dominant political economic trajectories, and what Code terms the "epistemologies of mastery" that propel teleological notions of development and progress (Code 2006, 4). *Extinction*'s focus on the irresolvable uncertainty of the relationship between the whole and the parts is crucial: global ecosys-tems and their functioning are understood as "known unknowns," and, as such, species losses and extinction cannot be apprehended through a calculative logic. Positivistic science, the Ehrlichs suggest, cannot distill in advance what species may or may not be expendable; this "refusal" is a point I return to in chapter 3.

An ecological universal and its "others"

There are perils in the Ehrlichs' analysis and vision, ones that con-tinue to underpin – and trouble – biodiversity politics to this day. In *Extinction*, the Ehrlichs argue that a global reduction in population growth will reduce the demand for fuelwood in the South and the consumption of commodities in the North, thereby helping prevent deforestation and species extinctions. This focus on population growth has a better-known antecedent in the Ehrlichs' *The Population Bomb* (1968), a neo-Malthusian work focused on population growth and scarcity. Outside their scholarly work, too, both Ehrlichs are long-time supporters of population control as environmental solution.[3] According to Connelly (2008), Paul Ehrlich at one point advocated cutting off food aid to countries that did not adopt contraception policies.[4]

Scholars have long noted the violence contained within the arguments regarding overpopulation. In a well-known 1974 essay, David Harvey wrote about the political connotations of the idea:

> The meaning can all too quickly be established. Somebody, somewhere, is redundant, and there is not enough to go round. Am I redundant? Of course not. Are you redundant? Of course not. So who is redundant? Of course it must be *them*. ([1974] 2001; emphasis added)

An ecological politics oriented around scarcity and overpopulation tends to mark some people as sacrifice-able in the name of all humanity. These people are often the othered "them" – those at a distance and who are not like us. In short, Harvey's observation is an acute reminder of the deplorable undercurrents within environmentalism, where only *certain* populations in *certain* places are made vulnerable to dispossession in the name of saving *Homo sapiens*.[5]

This is a continuation of distinctively liberal tendencies. Liberal ideals have always held universals like human freedom alongside ongoing or even deepening domination and oppression. Universal concepts of liberty developed and held high by John Locke and by American patriots sat alongside growth in slavery and colonialism (e.g. Losurdo 2014, Lowe 2014). Designations of difference as distance – the argument that "they" are not as evolved as "us" – was crucial in justifying colonial dispossession and slavery even when otherwise professing ideals of freedom.

Under an ecological universal, "they" are often deemed to lack the appropriate environmental ethic, perhaps "they" don't care enough about nature. These arguments are made in the name of saving the planet, but also justify authoritarian interventions from above, such as the imposition of a protected area. The Ehrlichs are an interesting case. They challenge liberal rationalities like the primacy of individual economic freedom while recreating new and old universals, this time of an ecological type. In the process they erect hierarchical differences between humans, between those who care about nature for its intrinsic *or* extrinsic value and those who don't.

Intrinsically bad humans (for the most part)

For the Ehrlichs, focusing on the benefits that nonhumans provide to human survival – the "extrinsic" value of species – is a kind of necessary evil; they themselves believe in the "intrinsic" value of nonhumans, or at least, the diversity of nonhumans. Unfortunately, though, according to the Ehrlichs, arguments about "the right to exist of nonhuman life forms" ("their aesthetic and intrinsic interest") or "appeals for compassion for what may be our only living companions in the universe" all "fall on deaf ears" (1981, 241). The Ehrlichs thus articulate a common belief in biodiversity politics: "most people" will only ever value nonhuman nature for its benefits to humans.

Others writing about the relationship between species diversity and ecosystem function all express ethical opposition to extinction rates, but lament that they are in the minority. Tropical biologist and creator of the "biodiversity hotspots" concept Norman Myers engages in this kind of self-exceptionalism in *A Wealth of Wild Species: Storehouse for Human Welfare*. He writes, "However much I may agree that every species has its own right to continued existence on our shared planet, I do not believe that the world yet works that way" (1983; cited in Takacs 1996, 35). The Ehrlichs, in *Extinction*, write, "Until ethical attitudes evolve further, conservation must be promoted as an issue of human well-being and, in the long run, survival" (1981, 241).

In short, the Ehrlichs suggest that it is not pragmatic to try to "save nature" for its intrinsic value because no one – aside from a few enlightened scientists and environmentalists – cares about it. This is the refrain that I heard uttered more than any other in the course of my research: "No one cares about nature." The implication, then, is that any intrinsic value must be – wearily, perhaps even despondently – set aside. Instead, biodiversity advocates should emphasize utilitarian or extrinsic value in order to appeal to the masses. As I described in the opening chapter, Pavan Sukhdev, head of The Economics of Ecosystems and Biodiversity (TEEB), a major international initiative to value global nature, said that the "necessity of economic valuation" for biodiversity conservation was "humanity's greatest failure" because it reflects the lack of an "ethic of care" for nonhumans. From this perspective, biodiversity advocates need to emphasize what biodiversity does for us, rationalizing these benefits via biophysical and/or economic calculations.

I empathize with ecologists and environmentalists' feelings of weariness in the struggle against biodiversity loss. At the same time, however, in light of the enormous complexity of human–nature relations in incredibly variable socio-natural circumstances around the world, I do not take as fact that there are these two simple ways to value nature. The choice is not between valuing nonhuman life and valuing only humans. Rather, this dualism of intrinsic versus extrinsic value – which has been produced by and continues to be central to much Western environmental thought and activism – provides a crucial understory to the whole enterprising nature project. Analysis of this dualism can help with understanding why Western environmentalism remains conservative and even, at times, reactionary.

Much critical scholarship has focused on the ways in which the "nature–culture" dualism animates modernity, originating in sixteenth-century colonial encounters and the emergence of an imperial global economy controlled by Europeans.[6] From this perspective, dualisms are not universal (although they may seem ahistorical and timeless), but rather have violent histories and geographies, marking some bodies and spaces as a part of nature and thus as less civilized, more exploit-able and/or in need of improvement. This other dualism – "intrinsic–extrinsic" – is also troublesome, particularly in the context of biodiversity conservation and the challenges of living with other species. As with the nature–culture dualism, this two-part worldview is not universal across cultures, epistemologies, and ontologies. Many scholars have suggested that a concern for species diversity is not a uniquely Western prerogative and, furthermore, cannot be captured within this dualistic frame of inherent worth versus utility. Anna Tsing writes about how, for the Meratus people in Borneo, survival is only one part of why they cultivate an incredible diversity of vegetables: "They value variety because of the taste, for the sociability it allows, for its sheer exuberance, and because it increases the chances of a bountiful harvest" (2005, 165). Larry Lohmann (1995) explains why Thai activists resist the dichotomies that form the basis of Western environmentalism, including what he refers to as the "ecocentric–anthropocentric" dichotomy, a dichotomy that is not relevant for rural Thai peoples' values and ways of living.

Yet, as a refrain in biodiversity conservation, and especially in a turn to enterprising nature, the intrinsic–extrinsic dualism is presented not as historical, but as a universal fact. Some care about nature, especially those who have evolved to this state of moral knowingness and ethical care, and the others, meanwhile, not only have failed to arrive, but also cannot even be trusted to ever make this enlightened state. There is a self-exceptionalism evident in Myers's and the Ehrlichs' statements, self-exceptionalism that typifies a strand of righteous Western environmentalism. Most humans are bad – they don't care – so we need to figure out scientific laws that can show "them" just how much we need nature. Universal scientific laws, this story goes, substitute in for politics and, especially, ethics (Latour 2004, Haraway 1988). David Takacs, in his book *The Idea of Biodiversity* makes a similar point based upon interviews with the ecologists at the center of biodiversity, such as E.O. Wilson and David Ehrenfeld. Takacs argues that the notion of biodiversity encompasses scientists' "factual, political and emotional arguments in defense of nature," including a belief in the "intrinsic value" of nonhuman nature, while "simultaneously appearing as a purely scientific, objective entity" (1996, 99).

To summarize, concerns with extinction and species loss are positioned as global biopolitical concerns, as crises that hold potential for enormous devastation alongside enormous uncertainty, crises that require precaution – conserving all the rivets. The promise of ecological thinking, then, emerges in a rejection of reductionism and narratives of mastery, progress, and growth – themes we see reflected in *Extinction*. But *Extinction* also directs us to the perils in Western ecological thinking, in seeking to impose universal laws defined in the Global North on land use and management elsewhere, especially in the tropical countries of the Global South. The rivet hypothesis disrupts narratives of mastery, of all-knowing calculation, but at the same time re-inscribes the call for a god trick, for universals that are severed from cultural, social, political, and material relations, from history and context. Universals that fill in, as the intrinsic–extrinsic dualism tells us, for the supposed immorality of (almost) all humanity.

The loss of biological diversity quickly transforms into a problem of global human concern. The rivet popping becomes caught up in geopolitics, national security, as well as visions of future profits, especially within the machinations of the US government. With the rise of the Convention on Biological Diversity, enterprising nature becomes crystallized in international environmental law.

Rise of the Crisis and Promise of Global Biodiversity

In his book *Saving Nature's Legacy*, historian of biodiversity Timothy Farnham (2007) credits ecologists Elliott Norse and Roger McManus with the first "official" definition of biological diversity. In the late 1970s and early 1980s, both were working at the US Council on Environmental Quality (CEQ), a federal organization that coordinates federal environmental efforts and works closely with agencies and White House offices in the development of environmental policies and initiatives. The CEQ's then-chair, James Speth (appointed by Jimmy Carter), requested that Norse and McManus research losses of animal and plant species and their habitats, and "write a new chapter for the next CEQ annual report on an unprecedented subject: the status of life on earth" (cited in Farnham 2007, 16).[7] Over the next decade, life on earth's "status" was identified as a crisis; US economic and imperial interventions proliferated as a result.

The 1980 CEQ report – *Ecology and Living Resources: Biological Diversity* – represented biodiversity as the aggregated population of species that underpins ecosystem functioning. In its general conception, Norse and McManus's framing parallels the Ehrlichs' rivet hypothesis.

At the same time, however, the authors also include *genetic* diversity: "the amount of genetic variability among individuals in a single species, whether the species exists as a single interbreeding group or a number of populations, strains, breeds, races or subspecies" (cited in Farnham 2007, 18). This definition shifts conservation from a focus on endangered species, wilderness, and extinction to a more encompassing project that includes genetic, species, and, in a subsequent report, ecological (community) diversity (Norse et al. 1986). The redefinition broadens biodiversity's focus from a concern for the health of individual species or subspecies to a concern with the diversity of life in general. This new focus encompasses all life, including insects, decomposers, and others that are not often captured through endangered species–focused conservation that centers on charismatic or else easily detected declining populations. In short, it is a shift toward "a totality whose implicit value underlies all arguments for individual species" (Takacs 1996, 56).

In the final CEQ report, Norse and McManus claim that biological diversity is humanity's "greatest natural resource, on which we depend for food, oxygen, medicines, psychological wellbeing, and countless other benefits" (cited in Farnham 2007, 18). A shift, here, is underway, in which US biodiversity advocates come to emphasize not only the utilitarian value but also the economic value of biodiversity. The report draws heavily from Norman Myers's 1979 book *The Sinking Ark*. Myers, a tropical biologist, focused not only on the risks to humanity from extinctions – the biopolitical use values – but also on direct economic benefits that species provide, both existing and future. These included economic benefits in the fields of agriculture, medical and pharmaceutical development, and industrial processes.

Myers also merges economic impacts with national concerns, discussing especially the "dearth of natural crop germ plasms" in the United States, noting that the country's crop production rests on a "very restricted genetic base" (1979, 63). What Myers is pointing to here is the relative "biological impoverishment" of the US in terms of genetic characteristics, noting that the country is "entirely dependent" on other countries to access new genetic characteristics (64). Wild strains of agricultural crops, Myers proposes, could be worth around $520 million per year in productivity gains (68), and this amount is dwarfed by the $3 billion annual value of plant medicinals (drugs derived from natural origins). Myers writes, "There can hardly be a more valuable stock of natural resources on earth than the planetary spectrum of species with their genetic reservoirs" (cited in Farnham 2007, 216).

Farnham suggests that Myers's utilitarian and economic arguments were crucial to building policy-maker support for addressing the problem of biodiversity loss (2007, 217). The array of government departments participating in the US Strategy Conference on Biological Diversity in 1981 – the first major event on biological diversity – demonstrated the wide range of interests galvanized by the new entity, biological diversity. As Farnham outlines, the conference was co-organized by the US Agency for International Development (USAID), the Department of State, the Department of Agriculture, the Department of Commerce, the Department of the Interior, the Council on Environmental Quality, the Smithsonian, the National Science Foundation, and the US Man and the Biosphere Program. However, despite involvement by this diverse set of departments, the conference focused on threats "to genetic diversity and losses to the agricultural and business sectors" (Farnham 2007, 221).[8] These governmental and scientific conversations culminated in the 1986 National Forum on Biodiversity organized by the National Academy of Sciences (NAS) and the Smithsonian. Hundreds attended the event, which attracted mainstream media attention and announced the crisis of biological diversity to the American public.

While the NAS wanted the National Forum on Biodiversity to be an "objective assessment of the problem of biodiversity loss," it ended up, at least for some, more like a "consciousness-raising event and media spectacle" (Takacs 1996, 39), as "biologists who loved biological diversity came out of the closet" (38). At the conference, a group of leading biologists – including Jared Diamond, Paul Ehrlich, Thomas Eisner, Evelyn Hutchinson, Ernst Mayr, Charles Michener, Harold Mooney, Peter Raven, and Edward O. Wilson – formed the Club of Earth, focused on preserving biological diversity. In *Biodiversity*, the edited book emerging from the conference, Wilson (the editor) argues that the focus on biodiversity at this time emerges out of the growing scientific data on deforestation and extinction rates, and the links between biodiversity loss and ecosystem health (Wilson 1988a).

Measuring biodiversity loss

But how do scientists like Wilson measure biodiversity loss? With such a broad definition encompassing not simply species diversity but also genetic and ecosystem diversity, biodiversity is challenging to measure. This is not surprising given that it is meant to encompass, literally, life on earth. In an acre of land or water (with some exceptions, of course), one could spend a lifetime and still not necessarily catalog all the living things, let alone their genetic diversity. To state the obvious, we simply cannot count all the species on the planet at any given time; we have to find ways to represent diversity, an ongoing and rich site of ecological research and debate.[9]

Given this lack of knowledge, and lack of baseline, how did the ecologists in the 1980s come to a consensus that species diversity (never mind the other axes of diversity – genetic and ecosystem) is decreasing over space and time? As Voosen (2011, n.p.) describes, "It is this lack of empirical data, the ebb and tide of unknown bugs, shrubs and grubs, that theoretical ecologists have attempted to fill with mathematics for the past century." Estimations of biodiversity loss in 1986 and today rely on the use of mathematical equations that allow for discrete quantitative measures of diversity to be produced. Estimates of species and rates of species loss are largely based on models of species–area relationships (SAR), which posit that the size (area) of a specific ecosystem type (say, tropical forest) is related to the number of species within it. Researchers in the 1800s, such as Alexander von Humboldt, noted these relations, which became concretized in the theory of island biogeography (MacArthur and Wilson 1967). Writing in 1988, Wilson posits that this relationship is geometric in island systems – the number of species "will increase as the fourth root of the land area" (1988b, 11). If land cover changes – such is sometimes the case with widespread deforestation and permanent alteration of the ecosystem – then the theory posits that the number of species will also decline accordingly. Thus estimates of species diversity changes are made based upon the changing size of the eco-system, all within a broader assumption of an equilibrium configuration or a norm from which ecosystems depart.

Myers, writing in the 1988 book *Biodiversity*, gives a sense of this mathematics in the tropics. Using remotely sensed data, Myers estimates that "remaining primary forests cover rather less than 9 million square kilometers, out of the 15 million or so that may have existed according to bioclimatic data" (Myers 1988a, 29). Drawing from theories of island biogeography, "which is supported by abundant and diversified evidence," Myers argues that "we can realistically expect that when a habitat has lost 90% of its extent, it will eventually lose half its species" (30). Drawing from examples in western Ecuador, Brazil's Atlantic coast, and Madagascar, all sites of major deforestation, Myers claims that the rates of extinction could be somewhere in the range of 1500 species per year, or 50 000 species in the last 35 years (30). Extending these kinds of cal-culations, Wilson estimates extinction rates in all tropical forests to be around 10 000 species per year (cited in Myers 1988a, 30).

An additional complication in quantifying the problem of biodiversity loss is that species extinctions are, as Myers (1988a) notes, a "fact of life" on planet earth. "Of all the species that have ever existed," he writes, "pos-sibly half a billion or more, there now remain only a few million" (28). However, scientists like Myers and Wilson compare present extinction rates with what is termed the "background rate" of extinctions. Drawing from historical ecology, Wilson estimates the rate of species loss at an average of one species a year. In the 1988 book *Biodiversity*, Myers suggests that present extinction rates are hundreds of times higher, or even thousands, and Wilson estimates present rates to be as high as 1000 to 10000 times the historical rate of extinction. This leads Wilson and others gathered around the Club of Earth to declare that we are living through the sixth great extinction (Wilson 1988b, 8; see also Myers 1988a), a term used to signify the current parallel with the five previous mass extinctions of life on planet earth. While there is widespread agreement that rates of diversity are decreasing, the calculations themselves – the scale and scope of the shifts – are a site of ongoing research and debate, especially estimates of global rates of loss.[10]

I present these challenges of measuring biodiversity loss here not to dispute "the sixth extinction" (see Kolbert 2014, Ceballos et al. 2015). Rather, my purpose here is to flag the enormity of the project of measuring biodiversity, the immensity of trying to capture the richness of the planet's diversity into single measures of extinction rates, past and present. Too, the species–area relationship is worth understanding as a productive mathematical equation that not only underpins the fact of biodiversity loss, but also directs courses of action by focusing attention on preventing land use changes, often through protected area creation. And it is a spatial relationship that hinges upon human-induced land use changes as inimical to biodiversity, a point that Vandana Shiva vociferously rejects (I return to Shiva's work in the conclusion of this chapter).[11]

Imperialistic natures?

E.O. Wilson's *Biodiversity* draws attention not only to facts and figures of loss, but also to the links between economic development and biodiversity loss. "Destruction of the natural environment is usually accompanied by short term profits and then rapid local economic decline," Wilson writes. There is, he goes onto say, "immense richness" in tropical biodiversity, which he says is an "untapped reservoir of new foods, pharmaceuticals, fibres, petroleum substitutes, and other products" (1988a, vi).[12] Similarly, in US governmental discussions, biodiversity became a consideration because it brought together scientific concerns but also the interests of multiple governmental bodies, as Farnham's (2007) book outlines. These interests related to the development and security of so-called "developing countries," which might be threatened by extinction and biodiversity loss as Wilson suggests; the US began to form ideas for intervention on this basis. For example, in 1983, based on the reports of the CEQ and other emerging evidence about the loss of genetic diversity, Congress passed the International Environment Protection Act, which directed the US government to assist developing countries in conserving "genetic resources." As a part of the process of the act's implementation, federal agencies were given the task of drafting a strategy for conserving biological diversity in foreign nations, directed by the USAID.

At the same time, US government bodies were interested in the opportunities for biotechnology and genomic industries to access a wide range of genetic resources for drug development and other applications, to access the "immense richness" of tropical forests, in Wilson's terms (1988a). This interest in genetic resources coincides with the growth of biotechnology. Melinda Cooper (2008, 4) notes that, beginning in the 1980s, the life sciences have "played a commanding role in America's strategies of economic and imperialist self-reinvention." The importance of the life sciences, she suggests, drove the US in its efforts to reorganize drug trade rules and intellectual property laws in its favor – and in favor of its industries in agribusiness, pharmaceuticals, and other biotechnology-based industries.

The rise of biodiversity as a global crisis is intimately linked to these efforts, to these winds of bioeconomic promise: a promise to save the economy, US hegemony, and *Homo sapiens*. In other words, biological diversity – as a global environmental crisis – has always been economic; it has always been linked to questions about value to the state and value as a potential and profitable resource. Biodiversity was not co-opted. Rather, its ascent into a global environmental concern has always been entangled with universal impulses to save humanity, along with economistic and imperialist rationales and practices. In this way, the rise of biodiversity conservation mirrors broader liberal tendencies to hold universal, higher order principles for the good of humanity (like political emancipation, social equality, scientific progress) alongside bald, imperial grabs for land and resources, in this case, grabs for genetic material.

Biodiversity's US-based germination – both within knowledge-producing institutions like Stanford and the governmental reports and committees described above – is critical to the rise of the concept as a global crisis of concern. It matters that the concept is consolidated in the US, supported by the political clout of the world's largest economy, military, and growing cadre of environment and development institutions and experts. But it would be narrow to say that biodiversity only emerges from or simply serves US interests, especially as it becomes debated at the United Nations and within international environmental law and policy.

The limitless potential of biodiversity

By the late 1980s, the economic impetus for biodiversity conservation was becoming institutionalized internationally. The 1987 Brundtland report *Our Common Future* represents biodiversity as a source of almost limitless wealth. The loss of biological diversity, the report states, is eroding the opportunities latent within diverse species and genetic resources, especially for "developing countries." Written at the start of the biotechnology boom, the report optimistically predicts that the economic value in the genetic resources alone "is enough to justify species preservation" (WCED 1987, 155). For example, the report suggests, wild species can increase agricultural efficiencies:

> More recently, a primitive species of maize was discovered in a montane forest of south-central Mexico ... The wild species is a perennial; all other forms of maize are annuals. Its cross breeding with commercial varieties of maize opens up the prospect that farmers could be spared the annual expense of ploughing and sowing, since the plant would grow again yearly of its own accord. The genetic benefits of this wild plant, discovered when no more than a few thousand last stalks remained, could total several thousand million dollars a year (155).

The report argues that – despite the rise of genetic engineering (GE) – wild genes are important. In fact, the rise of GE technologies does not render wild genes "useless" but rather positions them as "even more valuable and useful" because they provide unique characteristics that might not be found in increasingly uniform domesticated crops. Wild genetic material and unique genes could help conserve agricultural land, and even facilitate "harvesting crops from deserts, from seawater" (155).

While these quotes and others in the report emphasize individual species, the larger argument made in the Brundtland report is that we cannot know in advance which genetic material, within which species, could unleash an amount equaling "several thousand million dollars a year" (155). Biological information, the genetic resources of diverse species, is positioned here as a source of "green" economic growth.[13] Not all species hold the jackpot, but all hold the potential. The idea of this unknown future exchange value of biological diversity shares resonances with the Ehrlichs' rivet metaphor. Both ideas hold up extinction as potential loss or even disaster – a loss of future monetary value and a loss of necessary ecological function. Biodiversity loss holds a risk to both capitalist value accumulation and to *Homo sapiens*. This is a crucial synergy or alignment: biodiversity's biopolitical and exchange values are drawn together. We cannot distinguish between necessary and surplus species for ecological functioning, and so we must enact precautionary conservation; likewise, we cannot distinguish which species might yield benefits to the state or the market, so, again, we must enact precautionary conservation.

Thus, the Brundtland report's articulation with other circulating ideas reveals the emergence of the promissory nature of bioeconomies, their utopian impulse. The basic idea is that it is possible to align global social reproduction (value of nature to *Homo sapiens*) with national interests (value of nature to the state). Furthermore, at times, these global/national use values can also be reconciled with their capitalist value. Biodiversity then, at the global level, emerges not so much out of a challenge to the state and capital; rather, it is forged out of an uneasy alignment between state and capital, between these interests, rationalities, and global biopolitical and ethical concerns. In this new bioeconomic framing, ecosystem conservation – which is often understood as a constraint on economic growth – is positioned instead as a stimulus to national economic development, albeit managed, sustainable growth (Bernstein 2002).

These developing ideas of bioeconomies represent, in part, a pursuit of new opportunities for economic growth and the production of new sites for accumulation; in this sense, they reveal a neoliberal economic orientation. At the same time, however, and perhaps more important, this positioning of biodiversity resonates with Keynesian policies in the postwar era as a way to deal with uncertainties inherent to liberal capitalism more broadly, policies like state stimulus for unemployment or decent working class wages. Especially in *Our Common Future,* the biosphere is positioned as a key site for dealing with the uncertainties of the time. The biosphere, I argue, begins to occupy an important site not only for accumulation (as the source material for growing biotechnology applications) but also for protecting the advance of predominantly economic growth-centered development patterns.

Liberal Natures? The Rise of the Convention on Biological Diversity

Proponents of global biodiversity concern quickly became focused on creating an international mechanism or instrument that could place biodiversity conservation among the priorities of governments, both North and South. In 1990, the United Nations Environment Programme (UNEP) convened negotiations for an international agreement focused on biological diversity conservation. The first draft reflected views of the international conservation organizations and institutions located in the North. Particularly influential was *Conserving the World's Biodiversity,* a report produced by IUCN in 1990, in collaboration with four Washington-based organizations: World Resources Institute (WRI), Conservation International (CI), World Wildlife Fund (WWF) US, and the World Bank (McNeely et al. 1990). The IUCN report's authors include leading scientists and conservationists, all from the United States: Jeffrey McNeely, the chief scientist for IUCN, Walter Reid from WRI, and Russell Mittermeier, President of CI. All three of these scientists would go on to play critical roles in global biodiversity conservation; they are still active today.[14]

This report and the earlier, 1980 one demonstrate the spatial politics of conservation at this time (and still today). It is a mode of conservation that has its gaze fixed upon the lands of the Global South – spaces populated, in many cases, by peoples who have their own approaches to producing, protecting, and using diverse nonhuman species and are caught up in waves of colonial and neocolonial development projects. The circuits of biodiversity conservation are all in play here, dominated by Northern-educated experts working at UNEP, WWF, FAO, UNESCO, IUCN – expert managers of global "living resources." Maps in both of these reports clearly lay out priority conservation areas, the vast majority of which are in the South (see Figures 2.1 and 2.2). The 1990 report draws attention to "mega diverse" countries – those countries within which one finds high percentages of the world's biological diversity. Table 2.1 demonstrates the national calculus going on at this time, showing that even within the tropics there are nations who harbor more or less of the world's diversity. In this document, 12 mega diverse countries are identified, including Brazil, Columbia, Indonesia, and Mexico.

This IUCN report was heavily steeped in the economic rationales that made biodiversity loss an entity of concern with the US government; it focused on making an *economic* case for conservation, although less linked to the national interests of the US or biotechnology firms and more to diversity's role in green development. It drew heavily from McNeely's 1988 *Economics and Biological Diversity*, a publication that claimed policies for conservation must focus on showing the value of biodiversity "in economic terms" and, especially, be tied to a "country's social and economic development" (vii). Such a framing provided a correction to the 1980 World Conservation Strategy that was, as Steve Bernstein (2002) argues, panned by governments from the Global South for its lack of concern for developmental, poverty-reduction priorities in the South. Global policies, the McNeely report stated, must address not only the problem of biodiversity loss, but also, simultaneously, development issues and the unequal burdens of conservation. For McNeely, the costs of conservation too often fall to the "comparatively few who seldom are provided any economic incentive to conserve the resources" (1988, iv).

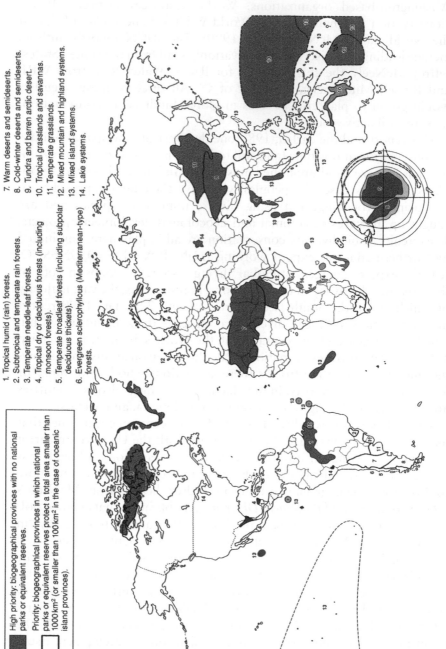

Figure 2.1 Priority regions for the establishment of protected areas, identified in the 1980 IUCN World Conservation Strategy. © IUCN, WWF, UNEP 1980 (by permission). From IUCN (1980).

High priority: biogeographical provinces with no national parks or equivalent reserves.

Priority: biogeographical provinces in which national parks or equivalent reserves protect a total area smaller than 1000km² (or smaller than 100km² in the case of oceanic island provinces).

1. Tropical humid (rain) forests.
2. Subtropical and temperate rain forests.
3. Temperate needle-leaf forests.
4. Tropical dry or deciduous forests (including monsoon forests).
5. Temperate broadleaf forests (including subpolar deciduous thickets).
6. Evergreen sclerophyllous (Mediterranean-type) forests.
7. Warm deserts and semideserts.
8. Cold-winter deserts and semideserts.
9. Tundra and barren arctic desert.
10. Tropical grasslands and savannas.
11. Temperate grasslands.
12. Mixed mountain and highland systems.
13. Mixed island systems.
14. Lake systems.

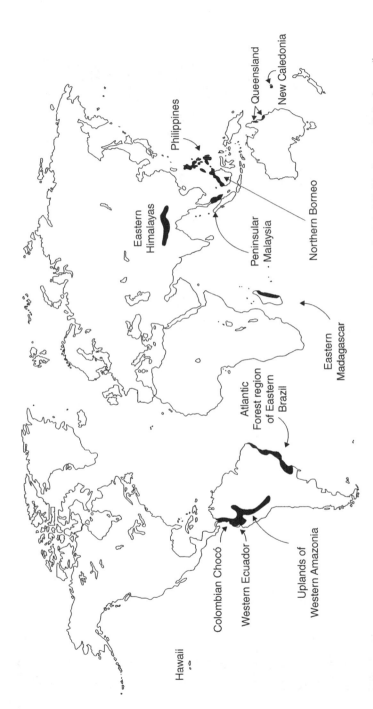

Figure 2.2 Biodiversity "hotspots" map from the 1990 IUCN report *Conserving the World's Biodiversity*. © Springer (by permission). From McNeely et al. (1990); original image from Myers (1988a).

Table 2.1 Countries with highest number of species for selected organisms, © World Resources Institute. (by permission). From McNeely et al. (1990, 89).

	MAMMALS	BIRDS	AMPHIBIANS
1.	Indonesia......(515)	Colombia......(1721)	Brazil......(516)
2.	Mexico.......(449)	Peru.......(1701)	Colombia......(407)
3.	Brazil.........(428)	Brazil......(1622)	Ecuador......(358)
4.	Zaire.........(409)	Indonesia......(1519)	Mexico......(282)
5.	China.........(394)	Ecuador......(1447)	Indonesia......(270)
6.	Peru.........(361)	Venezuela......(1275)	China......(265)
7.	Colombia......(359)	Bolivia......(±1250)	Peru......(251)
8.	India.........(350)	India......(1200)	Zaire......(216)
9.	Uganda......(311)	Malaysia......(±1200)	U.S.A......(205)
10.	Tanzania......(310)	China......(1195)	Venezuela/Australia......(197)

	REPTILES	SWALLOWTAIL BUTTERFLIES	ANGIO-SPERMS (est.)
1.	Mexico......(717)	Indonesia......(121)	Brazil......(55,000)
2.	Australia......(686)	China......(99–104)	Colombia......(45,000)
3.	Indonesia......(±600)	India......(77)	China......(27,000)
4.	Brazil......(467)	Brazil......(74)	Mexico......(25,000)
5.	India......(453)	Burma......(68)	Australia......(23,000)
6.	Colombia......(383)	Ecuador......(64)	S. Africa......(21,000)
7.	Ecuador......(345)	Colombia......(59)	Indonesia ......(20,000)
8.	Peru......(297)	Peru......(58–59)	Venezuela......(20,000)
9.	Malaysia......(294)	Malaysia......(54–56)	Peru......(20,000)
10.	Thailand/Papua NG......(282)	Mexico......(52)	U.S.S.R......(20,000)

Here McNeely is referring to the nations and communities of the Global South, who are being asked to conserve lands or resources, foreclosing other uses, foreclosing especially "development." The writers of these reports in the late 1980s and early 1990s, McNeely, Reid, and Mittermeier, are keenly aware of the "opportunity costs" of conserving tropical rainforests in the early 1990s, rainforests being cleared for the production of large-scale export-oriented crops and livestock. They are also keenly aware of the geopolitical context, where countries in the South are hostile toward ideas that they should devote more land to conservation, forgoing development.

The promise of a kind of bioeconomy – an economically oriented biodiversity conservation that could be an antidote to poverty and insecurity in the Global South – was widespread within initial discussions of the Convention on Biological Diversity (CBD) in the early 1990s. Pushed by governments from the South, the 1990 report produced by McNeely, Mittermeier, and Reid proposed that the international regime of the CBD would create economic incentives for conservation, particularly ones at the international level that would ensure that "wealthy nations benefiting from the biological resources of the tropics are able to invest in conserving the productive capacity of those resources" (McNeely et al. 1990, 15). The document reflects, to some extent, the (overzealous) enthusiasm of the 1987 Brundtland Report (*Our Common Future*), which saw the vast genetic resources of the tropics as an almost limitless source of wealth, wealth that could fund biodiversity conservation.

As an agreement on global biodiversity conservation advanced, the Group of 77 – the moniker for countries from the developing world, or the Global South – demanded that the CBD not only be a conservation agreement, but also focus on the problems of the appropriation of both knowledge and genetic resources – on biopiracy.[15] Since the vast majority of biotechnology firms are located in the North, the political economy associated with the rise of biodiversity is a familiar one – the resources are to be extracted from the "gene-rich" South (Shiva 1991, Cooper 1991), with benefits flowing to the technology- and capital-rich Global North. This situation was already well recognized by the time negotiations on the Convention on Biological Diversity began. In initial discussions of the CBD, a delegate from Malaysia told the UN General Assembly (cited in Cooper 1991, 111):

There are various instances where transnational corporations have exploited the rich genetic diversity of developing countries as a free resource for research and development. The products of such research are then patented and sold back to the developing countries at excessively high prices. This must cease. We must formulate mechanisms for effective cooperation with reciprocal benefits between biotechnologically rich developed countries and gene-rich developing countries.

In negotiations of the CBD, Southern governments (the Group of 77) demanded the creation of the international regulatory conditions that could allow them to realize their national enterprising nature – in other words, conditions that could facilitate the national extraction of value from bioprospecting. As negotiations continued, G77 countries, particularly the "megadiverse" countries, insisted on the sovereign rights of nations not only over territory, but also over the biological resources and the genetic information those resources contained. The preamble reads: "Reaffirming that States have sovereign rights over their own biological resources." This insistence on sovereign biodiversity was in direct contrast to the "common heritage of mankind" wording with which Northern governments and conservationists had begun. Furthermore, these demands led to the inclusion of what is known as the "third objective" of the CBD: to ensure the "fair and equitable sharing of benefits from the use of genetic resources" from "user" countries (i.e. those with biotechnology industries that use genetic resources, largely in the North) to "provider" countries (i.e. those countries with high levels of biodiversity, largely in the South). This objective is intended to stop biopiracy and ensure that the flow of monetary benefits from the commercialization of genetic resources could flow back to their "rightful" owners: the nations and communities of the global South.[16]

Kathleen McAfee (1999) aptly characterizes this strategy, with its focus on achieving international conservation and development simultaneously through the nationalization and commodification of genetic resources, as "green developmentalism." Green developmentalism describes an approach to economic development and environmental policy that entails bringing what is currently excluded from (and also harmed by) market relations inside these relations. Through the lens of green developmentalism, the globe's massive inequities and environmental problems are seen to be correctable by making the conservation of biodiverse spaces economic and, in the case of genetic resources, a source of economic growth and profit. Steven Bernstein (2002) terms this the compromise of liberal environmentalism, a compromise forged at the 1992 Rio Earth Summit that makes environmental concerns, economic growth and the market economy entirely compatible. At the time this strategy appeared to have some legs, especially as it came in the wake of a 10-year, $1.3 million bioprospecting deal between pharmaceutical giant Merck and the Costa Rican National Biodiversity Institute (INBio).[17]

Ecologists like the Ehrlichs might see the green developmentalist paradigm as pragmatic (albeit woefully) because it seeks to account for biodiversity's value within state interests. It gets around the problem that "no one cares," and also that the cards are stacked enormously against conservation in the Global South due to massive inequities and poverty.

Yet, green developmentalism, as McAfee (1999) argues in relation to the Convention on Biological Diversity, relies upon a discursive framing of nature as "an abstract, globalized resource, torn out of its spatial and socio-historical contexts" (137), one that "offers to nature the opportunity to earn its own right to survive in a world market economy" (134).[18] What this quote points to is the paradox at the heart of enterprising nature (past and present), in that it is a project that aims to render nature visible and legible for the state and capital while at the same time wanting to avoid the previous simplifications that backfired, such as scientific forestry and monoculture agriculture (Scott 1998). The approach aims to give nature the ability to prove its entrepreneurial value, but does so by once again wresting it from its geographical and historical contexts.

McAfee rightfully talks about green developmentalism as a bid or a play, an "aspiration" that is destined to fail, largely because it avoids addressing the underlying causes – "the existing economic structures or powerful state, individual or corporate interests" (150). Drawing from McAfee's still-classic essay, we may conjecture that what is going on in the rise of biological diversity, and the notion of the "bioeconomy" – from the Ehrlichs to the US government to the Brundtland to the CBD – is more than a techno/market fix, more than a compromise (as Bernstein (2002) suggests). It is rather, I contend, more like a utopian vision, in that what is being asked of the bios, of living ecologies, is to do something that appears to be not of our world. The world's biological diversity, the biosphere, is being asked to reconcile the needs of all humanity for healthy ecosystems (global social reproduction) with diverse national and firm interests (i.e. US interests in genetic resources, Costa Rica's interest in foreign capital to fund conservation, pharmaceutical giant Merck's interest in access to genetic resources), and overarching economic growth and development imperatives. It is utopian in the dictionary sense of the word, that it is a project that is, literally, no place: it is not of this earth, it is not of the socioecological systems that actually exist.

It takes an ecofeminist to put our feet on the ground.

Shiva Interruptus

A key disrupter of this view is scientist, environmentalist, and anti-globalization activist Vandana Shiva. Similar to organizations like GRAIN and Third World Network, what Arturo Escobar (1998) terms the "biodemocracy" collective, Shiva disputes the framing of biodiversity by organizations like the IUCN, the World Bank, and the World Wildlife Fund (WWF).[19] In her 1991 book *Ecology and the Politics of Survival*, Shiva argues that the "crisis of biodiversity erosion is focused as an exclusively tropical and Third World phenomenon." The "thinking and planning of biodiversity conservation," she continues, "is projected as a monopoly of institutes and agencies based in and controlled by the industrial North." She goes on to write that "it is as if the mind is in the North, the matter is in the South; the solution is in the North, the problems in the South" (1991, 7).

Shiva argues that it is Northern framings of conservation that characterize the Third World poor as "consumers" of biological diversity through their demand for firewood, fodder, and game meat. In fact, she argues, peasants and forest dwellers in the Global South actually *produce* biological diversity, achieving "production and conservation" simultaneously. The planet's biological diversity is not separate from cultural practices, Shiva argues. By arguing that peasants and farmers produce biological diversity, she challenges the notion of humans as being intrinsic destroyers. Shiva thus disrupts and interrupts the dualisms of "nature–culture" and "intrinsic–extrinsic" that are so prevalent in Western environmentalism. Her human exists in communities, it is a producer, a maker, a co-producer and consumer of life. Her human cannot live without a living soil, for example – a soil that can be made to be more alive through careful stewardship (Shiva 2014), as many farming communities demonstrate.[20] Her arguments recall Anna Tsing's (2005) point about the cultivation practices of the Meratus peoples: plants can be selected to build resilience, taste, and community exuberance. Shiva's vision, then, is one of decolonization: decolonization of not simply one nation over another, but a deeper sense of decolonization, decolonization of the conceits of Western notions of how to live, about what progress and development mean. In debates taking place at the constitution of the Convention on Biological Diversity, Shiva draws attention to the constant assertion of the superiority of Western systems of knowledge, calling for the respect and use of other knowledge systems with their own logic and epistemological foundations (Shiva 1991, 1997).

Shiva also disrupts and interrupts the circuitry of global biodiversity politics. She believes that the root of the biodiversity crisis lies with Northern institutions that promote "production systems based on uniformity and commodification," identifying in particular the World Bank's promotion of agricultural intensification (i.e. the "Green Revolution") that replaced crop diversity with crop uniformity. In particular, her 1997 text, *Biopiracy,* is a key moment of interruption into the rise of the global biodiversity apparatus. In this book, Shiva characterizes the rise of biodiversity as a continuation of imperialistic relations, rather than a challenge to them. Declarations of sovereign rights to genetic resources, she argues, are not a sign of Third World power, but rather a continuation of the colonial logic of John Locke and terra nullius. In *Biopiracy*, Shiva directly names the violences found within the bioeconomic dreams of the CBD, of the WCED, of IUCN, connecting these violences to longer histories of imperialism, dispossession, and control.

Shiva's politics, in the early 1990s and today, call attention to the role of Northern nations in fueling biodiversity loss. Rather than focusing on the land use practices of peasants in the Global South, she calls for more attention to ruling parties and elites who profit from land use change and monoculture, from export-oriented agriculture. These ideas of Shiva's, and of others who voice similar ones, however, are consistently marginalized within negotiations of the CBD.

Shiva is a complex figure. Scholars have criticized her for falsely representing subaltern voices (Jackson 1995) and for essentializing women in her variety of ecofeminism (Sturgeon 1997); journalists and columnists accuse her of overstating the impact of genetically modified foods (Specter 2014, Johnson 2014). These concerns, while important, do not negate Shiva's role as a disruptor of global biodiversity politics. Here I am particularly interested in her relationships to Western networks of power and knowledge; I find the contradictions in these relationships to be analytically productive in the context of enterprising nature. Shiva is internationally recognized and celebrated (the University of Victoria, where I worked when I wrote this text, awarded her an honorary doctorate), but her views are marginal to mainstream biodiversity discussions. Her views about Western liberal capitalism and the conceits of Western science are deemed radical, untenable, even perhaps naive and irrational, and yet she is internationally celebrated with awards and honorary doctorates; as one author described, "If [Shiva] personally accepted all the awards, degrees, and honors offered to her, she would have time for little else" (Specter 2014). Yet at CBD negotiations, views and politics like hers, expressed by organizations like GRAIN, ETC Group (formerly the Rural Advancement Foundation International), Third World Network, and Via Campesina are continually placed on the margins, seen to be making impossible, unrealistic demands. More recently, Shiva has opposed the rise of "enterprising nature" and the green economy, a turn which she says is "an attempt to technologize, financialize, privatize and commodify all of the earth's resources and living processes." She goes onto say, "This is the last contest between a life-destroying worldview of man's empire over earth and a life-protecting worldview of harmony with nature and recognition of the rights of Mother Earth" (Shiva 2012).

None of these statements is surprising for anyone familiar with Shiva and the politics of biodiversity. Her simultaneously revered and marginalized status is probably also unsurprising. This is a common tendency in multicultural liberal society, where race, gender, class, and geographical differences are celebrated, where accomplishments of those from the margins are feted regularly (i.e. a woman from the so-called Third World) so long as the person performs to the expectations or regulative ideals of liberal multiculturalism. But when the Third World woman refuses to respect central tenets or ideals of liberal society, she is cast out as deviant, silenced, or dismissed as "irrational."[21] In Shiva's case, she continues to be celebrated in the West, but her insights about historical and present inequities and about how to promote biological diversity are held at arm's length.

Yet, as I have argued in this chapter, it is the Western bioeconomic promise, exemplified by American scientists like McNeely, Mittermeier, or Reid and organizations like WWF, Conservation International, and IUCN, that might be viewed as hopelessly utopian, not Vandana Shiva. The so-called "pragmatists" hold onto an impossible dream wherein, once the conditions are right, all social, economic, and ecological values can be accounted for within a single analytical system – aligning global socioecological needs, national interests, and economic growth.

The promise of biodiversity is the ultimate realization of an "acultural vision": the Enlightenment promise of liberation by calculation. "Acultural" is a term of political theorist Charles Taylor (1995). It describes the tendency for so-called "moderns" to see their particular cultural manifestation as one of "convergence," as the one that all cultures naturally move toward (31). An acultural theory of modernity views Western liberal ideas and practices not as one specific cultural manifestation amongst others (with its own logics, rationalities, values, morals) but rather more like a "culture-neutral operation" (24). Such a theory of modernity is teleological: "human beings will just come to see that scientific thinking is valid, that instrumental rationality pays off, that religious beliefs involve unwarranted leaps, that facts and values are separate (Taylor 1995, 25). An acultural vision is one that removes the specificity of the Western configuration of logic and rationality, view-ing it as a universal project of human liberation, with all going in the same direction. And a major axis of this liberation is lodged in the Enlightenment powers of quantification and calculation, powers that promise to liberate humans from the uncertainty of future unknowns (Horkheimer and Adorno 1944), perhaps in this case, of ecological changes brought on by modernity itself.

Under the promise of biodiversity and enterprising nature, we are waiting, as the rest of this book charts, for the right measure, the right facts, the right models, the right laws and policies to realize humanity's (already known) destination.

Notes

1   The IUCN is considered the world authority for biodiversity conservation; it aims to alleviate extinction and preserve ecosystem function worldwide. Founded in 1948, it calls itself the world's largest global environmental network and includes more than 1000 government and NGO member organizations and almost 11 000 volunteer scientists in more than 160 countries. It is not an NGO, as it has governmental representatives.

2   Biopolitics, for Foucault, is a form of power that operates less around a Hobbesian "right to kill" or to "take life and let live," but rather through a desire to "promote life, to optimize the forces, aptitudes and life in general" (Youatt 2008, 401). This is a form of power that focuses on "administering and rationalizing life" (Youatt 2008, 400) to secure the health of the aggregate population. Under this form of power relation, violence and killing are rationalized only when they eliminate "biological threat[s]" to the species or race, or else improve it (Foucault 2003, 256). Foucault compares the operation of biopower with that of a more individualizing disciplinary power. For Foucault, biopower is "applied not to man-as-body but to living man, to man-as-living-being; ultimately, if you like, to man-as-species" (242). Its central concern is with the health of populations, and it operates via the aggregation of individuals, what Foucault calls "massifying."

3   See for example Vidal 2012.

4   Anne Ehrlich is on the board of the organization Federation for American Immigration Reform, which calls for immigration reductions and has been called a hate organization by the Southern Poverty Law Center. For more information see Beirich (2007).

5   For example, dispossession due to conservation interventions is well-documented as conservationists seek to "save nature" from a distance by imposing visions of pristine nature, emptied of humans, especially from places of privilege. See for example Brockington and Igoe (2006), Brockington (2002), Chatty and Colchester (2002). See www.justconservation.org for case studies of these politics in action (last accessed February 12, 2016). On neo-Malthusianism, geographer Tim Forsyth shows how science like the Ehrlichs' creates problems for farmers in the South whose livelihood practices – which do not affect the environment in the way the Ehrlich-type scientists suggest – are nevertheless restricted by policies based on this science (2002, chapter 2).

6   There is much to cite on this topic, but see for example Haraway (1992), Gregory (2001), Plumwood (1993), Latour (2004), Sundberg (2006), Braun (2002), Smith (1984), Harvey (1996). For an overview of the culture–nature dualism, see Sundberg and Dempsey (2009).

7   Speth would go on to found the influential World Resources Institute (WRI) in 1982, and WRI becomes a critical institution in promoting economic valuation approaches to biological diversity, and central to promoting an ecosystem service approach.

8   In its specific focus on genetic diversity, the conference was responding not only to the CEQ report but also to a 1978 National Research Council

report focused on the importance of conservation for their genetic material (germplasm resources).

9   The most commonly used measure is to count the number of different species in any patch of land, known as S – "species richness" – a measure that ecologists view as barely adequate. It is a measure that does not consider how many species (abundance) are in a particular location, nor their abundance in relation to other species (relative species abundance), which means it is a measure of the commonness or rarity of a particular species. It also does not consider evenness, which measures how close in numbers various species are in a given community (i.e. is there a similar number of different species in the area (even) or does one species dominate in numbers (uneven)). A more robust measurement of diversity is species diversity, which accounts for both richness and abundance. However, both species richness and species diversity only measure one part of "biodiversity" and there is more to biodiversity than species, including diversity within species (genetic diversity) and ecosystem diversity. To dip your feet into the academic research on biodiversity measures, see for example Alatalo (1981), Gotelli and Colwell (2001), Mendes et al. (2008), Chiarucci et al. (2011), Gorelick (2011), Leinster and Cobbold (2012), Tuomisto (2012), Dornelas et al. (2013).

10   Myers (1988a, 30) notes, "many reservations attend these calculations." Recent discussions about global measures include, for example, Butchart et al. (2004), Loh et al. (2005), Pereira and Cooper (2006), Collen et al. (2009), Lewis and Senior (2011), Scholes et al. (2008). Just as this book was going to press, a study by Gerardo Ceballos, Paul Ehrlich, and others used a higher "background rate" of extinction (two extinctions per hundred years, per 10 000 species) based on new evidence of mammal fossil records, and they also used records of actual vertebrate extinctions. This approach also yielded evidence that we are in the midst of "a global spasm of biodiversity loss" (Ceballos et al. 2015, 2), or that we are "entering the sixth extinction" (1).

11   See also more recently Mendenhall et al. (2012) (including co-authors Gretchen Daily and Paul Ehrlich) who question the assumption made in the species–area relationship that human-dominated landscapes are incapable of supporting biodiversity.

12   The book contains chapters by the "who's who" in ecology and within the main conservation institutions, including Paul Ehrlich, Norman Myers, Dan Janzen, Peter Vitousek, and Peter Raven. While dominated by ecological scientists, the book also contains chapters by economists, including one titled, "What Mainstream Economists Have To Say about the Value of Biodiversity," written by agricultural economist Alan Randall (1988). *Biodiversity* also contains contributions from specialists at the World Bank, the Smithsonian, and NGOs like The Nature Conservancy (TNC). The various networks and organizations that develop around the conference and book help to consolidate North America as one "center of calculation" for biodiversity. American academic institutions in biology are particularly central, but also important are conservation organizations and development bank institutions.

13   Many scholars have identified the role of genetic resources within a broader developmentalist paradigm. See, for example, Hayden (2003), McAfee (1999), and Castree (2003a).

14   For example, Reid went on to chair the Millennium Ecosystem Assessment, discussed in chapter 4. Mittermeier continues to be the president of CI.

15   At this time (and to this day), many Southern countries were facing biopiracy, which is "the appropriation of the knowledge and genetic resources of farming and Indigenous communities by individuals or institutions that seek exclusive monopoly control (patents or intellectual property) over these resources and knowledge" (ETC Group 2013).

16   The other two objectives of the CBD pertain to the conservation and sustainable use of biodiversity. As one legal scholar suggests, the "treaty balances the self-interest of one group of nations who have something they value and can offer to others, namely a rich supply of genetic resources, and the self-interest of another group of nations who have their own resource that they, in turn, can offer to the first group, such as technology and financing" (Tinker 1995, 194).

17   For more on this deal see Castree (2003a), Burtis (2008), Dalton (2004), Mateo et al. (2001).

18   I disagree with this catchall characterization of the articles and decisions of the CBD, which is more nuanced in its representations of nature and culture, and their relationality. For example, Article 8(j) of the Convention essentially links biodiversity and cultural diversity together, and has been a productive site of Indigenous organizing and achievements. More detailed investigations of the representations of nature/culture in international law and policy are needed.

19   Arturo Escobar (1998) helpfully organizes biodiversity politics into the "biodemocracy" collective, the globalocentric perspective of IUCN and associated Northern governments, the "Third World National perspective" camp of the governments of the Global South, and the social movement collective that comprises autonomous, place-based movements.

20   Shiva (2014) writes: "We are the soil. We are the earth. What we do to the soil, we do to ourselves. And it is no accident that the words 'humus' and 'humans' have the same roots."

21   Shiva has been called anti-science, especially her views on genetically modified organisms (GMOs) and agriculture. A *New Yorker* profile in 2014 charged her with overstating the impact of GMOs on farmers as well as claims about the prices of GMO seeds (among other things). The writer, Michael Specter (2014), drew heavily on pro-GMO advocates like Mark Lynas, who said of Shiva: "She is very canny about how she uses her power ... But on a fundamental level she is a demagogue who opposes the universal values of the Enlightenment."

3

An Ecological-Economic Tribunal for (Nonhuman) Life

Genetic resources from tropical forests proved to be no "gene gold." Bioprospecting and the gene rush did not lead to widespread conservation in the Global South. The future, unknown profit from genetic resources could not overwhelm the present, known profit from traditional commodity production, especially agricultural commodities like soy and sugar cane.[1] Viewed 20 years after the agreement of the Convention on Biological Diversity (CBD), a promise of "green developmentalism" based on upon genetic resources seems naive, almost ridiculous. But even the most active promoters of "gene gold" as conservation strategy voiced concern over its inadequacy well *before* the CBD negotiations concluded in 1992 – even though, as I previously noted, the agreement hinged on assigning strong national property rights for genetic resources.[2]

Writing in 1988, Jeffrey McNeely and his colleagues at IUCN advocated the calculation of what they called the "indirect values" of biodiversity, on top of the so-called direct values like those of genetic resources. Indirect values are those ecosystem functions and services that flow from biological diversity. For McNeely, these services – such as carbon sequestration, water purification, or pollination – when calculated, "may far outweigh" the values of genetic resources (1988, 19; see also McNeely et al. 1990). Valuation of these services, he argued, would alter land use decisions in the Global South that were responsible for transforming diverse ecosystems (including agro-ecological ones)

Enterprising Nature: Economics, Markets, and Finance in Global Biodiversity Politics, First Edition. Jessica Dempsey.

into monoculture land uses for soy, beef, and sugar cane production – land uses that were (and remain) a driving force of biodiversity loss.

But how does one quantitatively value the work of biodiversity of ecosystems? This chapter examines the articulation between two knowledge systems – ecology and economics – as biodiversity advocates begin work on the project McNeely and his peers laid out: calculating the indirect values of ecosystem services in order to make conservation the economically rational choice.

An Ecological-Economic Tribunal

"To effect a true synthesis of economics and ecology is the second most important task of our generation, next to avoiding nuclear war," wrote Robert Costanza and Herman Daly, two of the first "ecological economists" (1987, 7). That these writers placed the economics-ecology synthesis second in importance only to "avoiding nuclear war" verges on comical hubris, the elevation of their own work to absolute global necessity. Yet, many conservationists and scientists deem such a synthesis to be necessary for an environmental problem to even become a *real* problem: economic costs and benefits are what will propel an issue like biodiversity loss into the sights of state or market concern.

In the early 1990s, the Beijer Institute of Ecological Economics was a key site where ecologists and economists came to mutual understandings, and began building a true synthesis needed to make biodiversity loss legible and calculable for policy, to make the problem "real" for policy-makers. Collaborations at the Beijer created a new disciplinary apparatus, the "economics of biodiversity," enabling the problem of biodiversity to be represented via supply and demand curves. The Beijer Institute is a rich site to observe ecologists and economists grappling with how to turn the poorly understood and dynamic relations of ecosystems into entities – into "units" even – that can be judged by an ecological-economic tribunal.[3]

What is an ecological-economic tribunal? A tribunal, in general terms, is called to adjudicate on some issue or concern; it tends to be composed of experts or noteworthy public members tasked with making a judgment. An ecological-economic tribunal, however, aims to adjudicate through measurement and quantification, a form of judgment requiring the production of competitive, enterprising subjects. The Beijer Institute is a site where researchers wrestle with the framework, methodologies, and policies that can bring into being an ecological-economic tribunal for nonhuman life on earth.

Within the Beijer, we encounter a politics of measure, a politics aptly defined by Geoff Mann as "persistent efforts to articulate the unity of quality and quantity" (2007, 27). Here I am interested in the Beijer project's efforts to render the unruly qualities of ecosystem dynamics (and the enormously different living beings that comprise them) quantitative so they could be compared, ranked, and ordered. This kind of rendering, these "persistent efforts" as Mann and many others before him, from Marx to Horkheimer and Adorno have argued, is a central dynamic producing hierarchical value in capitalist societies – value in exchange, but also value to the state, perhaps in aiding the allocation of tax revenue or to inform governmental resource or land use planning.

Martin Weitzman, a Harvard economist participating in the Beijer project on biodiversity, explains the need for calculation as critical to achieving efficient allocations of scarce resources. For Weitzman, biodiversity conservation is a zero sum game and, given limited resources, "preservation of diversity in one context can only be accomplished at some real opportunity cost in terms of well-being forgone in other spheres of life, including, possibly, a loss of diversity somewhere else in the system" (Weitzman 1995, 21). If conservation is a question of rationalized trade-offs, then certain ecological elements must be prioritized. As this suggests, certain species, assemblages of species, relations between species (including humans), are more or less enterprising than others; they contribute more or less to the health and well-being of the state or firm, or community, they are more or less worthy of investment. At Beijer, the economics of biodiversity began to create the conditions – the right epistemological-ontological frameworks – to demarcate *quantitative difference* between the enterprising capacities of nature.

The economics of biodiversity is not necessarily a first step toward new commodifications. But it is always trying to produce a tractable biodiversity, tractable for governance, no matter the specific breed of policy: from social democratic to "command and control" to fully market oriented. As Weitzman states, if you cannot "define or measure diversity, then it is difficult to speak meaningfully about the best way to preserve it" – the "best way" being the most economically efficient approach (1995, 166). And economically efficient conservation, I suggest, is code for what I believe is most sought in the grand synthesis of economics and ecology: a way to govern ecosystems that is neutral and objective, not only scientifically, but socially. According to this logic, without quantity to determine the value of biodiversity to the state, to the market, or even to individuals, all social decisions are destined to be adrift in the tainted waters of subjectivity.

The Beijer project on biodiversity sought to produce the conditions for a kind of "god trick," to use Donna Haraway's term (1988). A god trick is Haraway's critique of objectivity in science, in that it presents itself as synoptic and impartial, a "view from nowhere." In the Beijer project, scientific objectivity is combined with economic objectivity to produce a god trick that seeks "infinite vision" not only of ecosystems and the creatures that make them up, but of their usefulness. It desires not only universal fact, but also universal value, a value produced so objectively that human subjects are only present as passive conveyors of the ecological-economic facts of life. This, I argue, is how we might picture an idealized ecological-economic tribunal: as a desire for a kind of supreme god trick that can adjudicate from a distance, combining ecological and economic facts of life.

This chapter is about the rise of this "true synthesis" that Costanza and Daly think is akin to avoiding nuclear war, focusing on biodiversity. It is about attempts to find the "best way" to represent the monoculturization of the planet, pointing us to the most objective, and thus right, interventions that can stop it. And so the Beijer Institute reflects a well-known play to determine the right way to know, live, and be from the center, from Northern-based experts. Universalizing god tricks in the service of saving nature are never innocent, emerging from on high to dictate how and where people are supposed to live, marking some humans and nonhumans as more or less worthy of investment, or even "killable."[4]

Ecologists and Economists Unite!

> I found myself to be at Stanford the same time as Partha Dasgupta. And that led to ... these evening get-togethers for the first time, on campus, of people in ecology and evolution, and economics. Ken Arrow was [also] there, David Sterritt, I ended up working with both of them quite a bit. Larry Goulder was a young new faculty member. ... This was a fun new thing.
> I got to know them personally. ... I was an early graduate student at the time, and ... this never became a focus of my PhD work, but I started getting to know some of the issues and some of the people. (Gretchen Daily, interview)

Ecologist Gretchen Daily is best known for her role in ushering in the ecosystem service revolution with her 1997 book *Nature's Services*. In the foregoing quotes, though, she outlines the networks and relationships forming in the late 1980s and early 1990s between the ecologists and economists concerned with biodiversity loss. These personal meetings at Stanford, as Daily went on to explain, were critical in getting past what was "a real war going on between ecology and economics." The "war" Daily identifies was epitomized by the debate between Julian Simon – an economist – and Paul Ehrlich (Daily's supervisor at Stanford). The "Simon–Ehrlich wager," made in 1980, had Ehrlich betting that the prices of five commodities would rise in light of resource scarcity due to population growth. Meanwhile, Simon bet that these prices would decline due to substitutions and technological innovations. At the time, the Ehrlichs held economists in disdain, seeing them as perpetuating notions of limitless growth and substitution on a bounded planet.

Speaking generally, Northern environmentalists in the 1960s–1980s expressed concern with the limitless growth, culminating with the Club of Rome 1972 *Limits to Growth* report (Meadows et al. 1972), which argued that both population and industrialization would hit the ceiling of the earth's system capacities. They went on to suggest a need for a global plan that could facilitate the transition to a steady-state equilibrium. Economists shot back that ecologists were alarmist, proposing far too restrictive constraints on economic development as well as the "natural" human ingenuity that would overcome such limits. Such a debate mirrors the perennial debate in liberal societies about the right amount of government intervention: not too much to constrain "ingenuity," but not too little to descend into "chaos." (It also reflects a newer dilemma: how to account for unknowns in liberal governance, a subject I return to later in this chapter and throughout the book.)

As Daily explained to me in a 2010 phone interview, it was discussions with eminent economists like Partha Dasgupta at Cambridge and Nobel Prize winner Kenneth Arrow at Stanford that helped her appreciate what economists did and how, as she describes, "superficial" many of the "ecologists' warnings were ... saying that from an ethical point of view – and this is slightly oversimplified – the only thing we can do is save nature." In engaging with economists like Dasgupta, she began to see that "there are deep ethical issues from many different perspectives." She described how ecologists and economics came to better empathize with one another:

> It made me realize first of all that many of them cared just as deeply as the biologists I knew cared about the natural world, and the future of human well being ... they made me appreciate how carefully they had studied a lot of these things ... So I slowly came to appreciate what economists can do and what they think about and how useful some of their approaches are. It is not just slapping a price tag on something, but they really have huge sub-disciplines on the effects on different sectors of society, or different types of people in a household of alternative policies. They thought ... about much more aspects of environmental problems that I have ever heard of from anybody else. It was clear that we needed to team up. They

[the economists] were also missing things ... they didn't know the science very well. And they were skeptical of some of the science because of the way it had been presented in the media. But as they got to know us, they were respectful. And soon we became all one group.

For Daily, these personal meetings allowed these academics to build "a common set of values," to get past their divides. Many of the academics who met in Stanford became involved in the Beijer Institute.

The Beijer Institute Project on Biodiversity

Established in 1977 as an Institute of the Royal Swedish Academy of Science, the Beijer Institute reorganized in 1991 to focus on ecological economics, with an aim to promote a deeper understanding of the interplay between ecological systems and social and economic development. It has been a critical meeting point for various ecologists and economists, enabling connections that are often described by participants as crucial in "overcoming critical barriers to communication between the two disciplines: rigidness in thinking and lack of confidence in the other discipline's perspective" (Barbier et al. 1994, xi). In an interview, Daily described the community around the Beijer as a "really alive community, and full of passionate and intelligent people, working on really the ways people depend on nature, and the institutions that might help protect nature, and human livelihoods and such." As the Costanza and Daily view about the desirability of an economics-ecology synthesis suggests, it seemed common sense to participants that these are the two critical academic disciplines that must communicate and come to some kind of consensus with each other in order to comprehensively picture, enumerate, and ultimately solve the biodiversity problem.

Alliances with economists can be a good strategic move, given that they often act less like social scientists and more like "experts who give advice to decision makers," as Michel Callon explains (in Barry and Slater 2002, 300). This cozy relationship is helped along because economics is often viewed as a knowledge system with the strongest "laws" in the social sphere, akin to physics in the physical sciences. As "the most abstract and mathematical of the social sciences," Timothy Mitchell writes, economics has "claimed the task of representing what seemed the most real aspect of the world" (2002, 82). Economic representations bring environmental issues *to life* in capitalist markets and states, which is not often the case for ecologists, who, despite their privileged position, barely register in the hierarchy of power and influence.

The first project proposed by the Beijer participants focused on the economics and ecology of biodiversity loss, the topic of the annual meetings in 1991 and 1992. This focal area was "motivated by the problem of extinction" but had a "broader agenda" in that it meant to address the "driving forces behind all socially undesirable change in the composition of species" (Perrings 1995a, xi).[5] The project lasted only five years – from the first meeting to research dissemination – but under the helm of economist Charles Perrings, the project managed to define the problem, identify gaps in knowledge, and concretize a dominant narrative about what should be done. Participants included the now "who's who" of this area: ecologists and biologists like Paul Ehrlich, Gretchen Daily, Brian Walker, Joan (née Jonathan) Roughgarden, and Crawford Stanley "Buzz" Holling; members of the discipline of ecological-economics like Robert Costanza and Carl Folke; and more "mainstream" environmental economists like Perrings, Partha Dasgupta, Timothy Swanson, Edward Barbier, David Pearce, and Kenneth Arrow, many of whom were prominent not only as academics but also as advisors to governments and international institutions. Other participants, important because of links to policy and practice, included Jeffrey McNeely, the chief scientist from IUCN, as well as representatives from the World Bank.[6] The goals of the Beijer biodiversity project exemplified the modern model of technocratic policy making, and the role and power of experts: "Our task is to improve our understanding of this problem and to provide guidance to policy makers on how to begin devising solutions" (Perrings 1995c, xiii).

The group convened only two conferences: one in September 1991 and the other in July 1992. At the first conference, the group identified four research themes;[7] at the second conference, groups presented research on those themes. The project was prolific in output, including two special editions of *AMBIO*, special editions of both *Environmental and Resource Economics* and *Ecological Economics*, and no less than three books (Barbier et al. 1994; Perrings et al. 1995a, 1995b).

Ecologists and economists collaborated on several articles, and these articles are filled with mathematical equations, mathematics being an important mutual "language" shared between the two disciplines.[8] Summarizing the research conducted through the project, Barbier et al. conclude: "Substantial progress appeared to have been made in opening lines of communication between the two disciplines ... The emergence of many points of common ground between ecology and economics on the analysis of biodiversity can be found" (1994, xii). In other words, the project on biodiversity was successful, at least in part, in translating between the disciplines of ecology and economics.[9]

Biological sciences and economics are not entirely new bedfellows, having met before in the forging of various natural resource management sciences and economics, from fisheries to rangeland management to forestry. This particular meeting diverges in that ecology in general, and the ecologists involved at the Beijer in particular, are themselves critical of resource sciences for their simplicity and exclusions.[10] Many of the ecologists participating in the Beijer project spent their careers responding to the *failures* of reductionist, single-use commodity models like those used in fisheries and forestry, which sidelined the complexities of ecological systems. Drawing from Buzz Holling's resiliency model, the ontology of living systems is less about stocks and stability, and all about dynamism and uncertainty.

Despite these differences, there are commonalities between resource economics and the economics of biodiversity, in that the economic theory remains relatively intact, emphasizing utility, and especially in the drive to identify points of equilibrium between economic growth and some "level of biodiversity."

Finding the sweet spot for conservation

In an introductory chapter to one of the three books produced by the Beijer project, economist Charles Perrings and ecologist Hans Opschoor describe the problem of biodiversity:

> This may be referred to as the problem of biological diversity conservation. It requires neither the preservation of all species, nor the maintenance of the environmental status quo. Instead it requires the development of the informational, institutional and economic conditions in which the private use of environmental resources will be sustainable – where sustainability implies the maintenance of a level of biological diversity and a scale of economic activity that will guarantee the resilience of the ecosystems which support human consumption and production. (Perrings and Opschoor 1994, 1–2)

The second sentence provides a somewhat startling frame for a new research agenda; the problem of biodiversity, Perrings and Opschoor write, does not require the "preservation of all species." But neither, they say, does it mean that the status quo should continue. The problem of biodiversity, they propose, involves identifying a sweet spot in planetary conditions, an equilibrium where economic activities and the "right level" of biological diversity align.

The authors associated with the Beijer project pay a great deal of attention to market failures, to the *non-equilibrium* state of global political economies where the "market value of biodiversity loss" does not "measure the change in social welfare associated with that loss" (Perrings et al. 1992, 202).[11] At the root of this state of affairs, as economists like to say, are the "independent decisions of the billions of individual users of environmental resources worldwide" (Perrings et al. 1992, 205). These decisions, they contend, "have been privately rational, given the information available to the decision-maker" (204) but fail to achieve the best outcome for society. This non-equilibrium situation is otherwise known as a classic externality situation, where private and social values diverge. Economist Timothy Swanson represents the "global biodiversity problem" as a supply and demand curve (see Figure 3.1), arguing that billions of instantaneous, decentralized decisions all over the globe favor land use change, what he describes as an ongoing process of replacing naturally occurring species with what he calls "human-chosen ones" (1995, 226), such as through agriculture or forestry.[12] The notion of externalities is an idea central and long established in environmental economics (e.g. Coase 1960); in the Beijer, we see the economic theorem mobilized with reference to the problem of biodiversity loss and brought to the attention of a wide range of ecologists.

Swanson's graph models the increasing global land conversion over time, with the horizontal axis representing the "passage of time" and the conversion of the "state's naturally evolved slate of resources" to uses like "modern agriculture" because of the "net benefit"

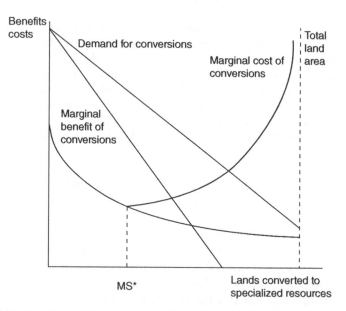

Figure 3.1   Supply and demand curve for optimal biodiversity policy. © Cambridge University Press 1995 (by permission). From Swanson (1995, 231).

states receive from this conversion (Swanson 1995, 229). The final point on the right represents complete conversion, a process fueled by these ever-increasing national benefits (this is represented by the increasing area above the "marginal benefit of conversions" line over time). [13] Here I want to draw attention to the way Swanson represents the problem of global biodiversity loss and its solution. The externality is demonstrated by the upward sloping "marginal cost of conversions" curve, which increases gradually at first, and then more rapidly as conversions increase. For Swanson, this line represents the global costs from land conversion (and biodiversity loss), costs that stem from "lost global services" such as "irreplaceable insurance and information services" (1995, 233) that biodiversity provides. This curve has multiple layers beneath it; it derives from ecology as it is a representation of the species–area relationship discussed in the previous chapter, which posits that increased loss of habitat (from conversion) leads to reductions in species numbers, in a log-linear fashion. The "cost" of the conversion of land, however, is one that is represented without hard numbers, largely because such estimates do not exist. These rising costs (even if we don't know these costs in any specific form, globally) are the externality: the costs represented by this curve are unaccounted for, or invisible; because "the opportunity costs of conversions in terms of lost global services are not included within the converting state's decision-making framework" (233), they are not included in the *national* and *individual* decisions that, for Swanson, drive land conversions and biodiversity loss. The problem of biodiversity loss then originates in this divergence between those who (supposedly) benefit from conversions – the state and consumers – and those who lose from the conversions – the planet or global humanity.

In an ideal world, where these costs of species loss are accounted for, the conversion process would halt at the intersection between the "marginal cost" curve and the "marginal benefit" curve, at point MS*, or marginal state. This is the promised land, the equilibrium point, the sweet spot of economists! This is the point where the benefits of land conversion are equal to the costs of conversion. If we stop just a small bit before that spot, it is less-than-optimal benefit (to the nation, to individuals); if we go a small bit after that spot, it is too much cost (to the global). This boundary setting requires a mode of calculation and governance that aims to both identify the MS* point and design policy interventions that can bring externalities inside the market, to make states and individual consumers "recognize the *true* opportunity costs of diverse resource conversions" (Swanson 1995, 252; my emphasis). But how do researchers know what the costs of conversion are? How do they calculate them? How does biodiversity factor into this? These are the problems that the Beijer project grapples with, problems that are by no means solved.

Before we move to Beijer's responses to these questions, however, I want to summarize how the problem of biodiversity loss is produced through the Beijer. It is a problem that takes place – to draw from Mitchell – in the sphere of "locationless" exchange, a problem "formulated geometrically, by the axes of a chart, as the two-dimensional plane" (2007, 85). The word "locationless" is crucial, since the chart is nowhere but everywhere at the same time, a key marker of what Donna Haraway describes as the "god trick," a universally applicable representation of the world, from a point of view so far above the earth it is "nowhere." In this graph, ecological devastation is produced as the result of "inanimate factors and faceless forces," by the natural actions of sovereign nations and fueled by the billion of decisions of "humanity," a point made nicely by Kathleen McAfee's 1999 article "Selling Nature to Save It?" Swanson uses the term "human-chosen" life forms in explaining biodiversity loss, but which humans – where and when – decide what nonhuman species to propagate or remove from the land? Under what terms? And what about that "humans" do not exist in opposition to biodiversity? What about humans as producers of nonhuman diversity, a central point made by Vandana Shiva? In this way, this representation is not simply technical; it is wholly political, by the way the representations exclude.

With its undifferentiated "humanity" making destructive choices, this approach to biodiversity loss effaces specific geographies and histories of biodiversity loss, obscuring colonial and neocolonial exploitation and the uneven flows of benefits resulting from this exploitation. The two-dimensional frame excels at framing out violence of the specific institutions and development agencies that fund and promote the economic activities that fuel biodiversity destruction and dispossession – activities powerfully described by Vandana Shiva and others in their 1991 book *Biodiversity* (Shiva et al. 1991). This frame does not reference the enormously uneven purchasing power of individuals and states that means that the billions of "autonomous" decisions are not even close to evenly located throughout the planet or across bodies, but are deeply inflected by individuals' locations in city or country, North or South, and by race and class and gender. In this locationless and ahistorical view, biodiversity loss is removed from colonial-capitalist processes, what Ann Stoler (2008, 2013) describes as "processes of ruination": the discursive-material processes of annihilation, displacement, and replacement driven by ongoing imperialist relations.

Biodiversity loss, in the Beijer view, is a technical problem that can be fixed by finding a self-sustaining equilibrium between the demands of consumers, national development interests, and the supply of global benefits stemming from biodiversity. In representing the "nonmarketed" world and especially what it lacks, Mitchell argues: "economics tends to diagnose ... defects as an absence of the techniques of *representation*. Things are stuck outside the market because they are not properly represented – by property records, prices, or other systems of reference" (2007, 248; emphasis added). The problem is not one of violence or relations of power, but rather one of a lack of proper representation that can be solved by the right way of marshaling, or finding unity between, quality and quantity.

In part, then, the Beijer project is animated by the search for the holy grail of equilibrium: finding and fixing "the *appropriate* level of exploitation – as well as conservation – of our natural environment" (Barbier et al. 1994, 18), or the most efficient *level of biological diversity*. This drive for equilibrium, this interest in defining the right balance between development and conservation makes the economics of biodiversity remarkably similar to the "old" resource economics.[14] The desired ecological-economic tribunal is one that governs objectively and neutrally through quantification, through measurement, from a distance.

The "economics of biodiversity" emerging from Beijer in the early- to mid-1990s demonstrated a range of policy approaches. Swanson's graph, for example, demonstrates a dream of efficiency and harmony that Friedrich Hayek called "catallaxy": a peaceful world order that would result from a perfectly operating market. Some writing out of the Beijer program focused on property rights, advocating an increase in the intellectual property rights of countries in the South over biological resources, as well as the development of an international system of property rights in national genetic material to incentivize conservation (Swanson 1995). Others called for the creation of international payments or transfers for tropical forests or biodiversity conservation areas, foreshadowing the present-day Reducing Emissions from Deforestation and forest Degradation (REDD) initiative, a UN mechanism that aims to pay countries for the carbon in standing forests (see Barbier and Rauscher 1995). One of the most comprehensive suggestions is a system of "Tradable Development Permits," a kind of cap-and-trade system for land use.[15]

But more command-and-control approaches can be found in the Beijer work as well, including discussions about the need for safe minimum standards meant to ensure biodiversity loss does not exceed ecological thresholds and cause permanent shifts in ecosystems.[16] A safe minimum standard would seek to "bound the level of economic activity in a way that will minimize the risk of irreversible damage" (Pearce and Perrings 1994, 5).

The idea of limits and thresholds (represented by safe minimum standards) was not uncontroversial within the Beijer Institute, particularly in relation to the economic constraints of such limits. Some participants clearly argued that the imposition of safe minimum standards must be the first step to "preserve threshold limits" (Barbier et al. 1994, 189), and the use of traditional regulatory and market-based instruments could be used to optimize allocations within these limits. However, others within the project made a counterargument, stating that the need for limits and thresholds can "be overly exaggerated in many cases" (189), and that the problem of biodiversity loss may be eliminated by simply eliminating market and policy failures, particularly government subsidies. In this view, limit- and threshold-based approaches are policies of last resort, but they are not excluded from the frame entirely.[17]

Common to both these approaches – tradable permits and safe minimum standards – is a need to find a way to make these "tractable," to make the diverse living things and the ecological processes they produce measurable to inform state and market decisions.[18] The Beijer participants note over and over again how biodiversity lacks the conditions, the systems of representation (price, models, laws, and policies) needed to find equilibrium, the MS*, on any scale. Achieving neutral valuation requires the rendering of quality into quantity. The needed "set of measures" for Perrings et al. (1992, 202) are those that can establish "the economic value of a change in the level of ecological services associated with a change in the level of biodiversity" (204). In other words, the measures needed are ones that would compute the change in the ecosystem services – say in carbon sequestration or water purification – that occur given a change in, say, species distribution, richness, or abundance.

This begs the question: what kind of ecological model can support such a demand?

Ecology in the Ecological-Economic Tribunal

The Beijer project sits upon the resiliency theory of Buzz Holling (1987, 1973), also a participant. This theory views ecosystems as dynamic and uncertain, at times highly stable and at others highly unstable. The notion of resilience is a part of a larger change in ecological thinking from earlier models of ecological change – namely the "Clementsian" view of ecosystem succession, which suggested that ecosystems go through highly ordered steps that lead to a climax assemblage of species. In resiliency-based ecological understandings, ecosystems do not have a single equilibrium state; change is viewed as episodic with sudden reorganization in relation to rare events and multiple possible states of "equilibria" (Holling et al. 1995). Such an understanding is at cross-purposes with much of resource science like forestry and fisheries which assume stability or, at least, predictable patterns.

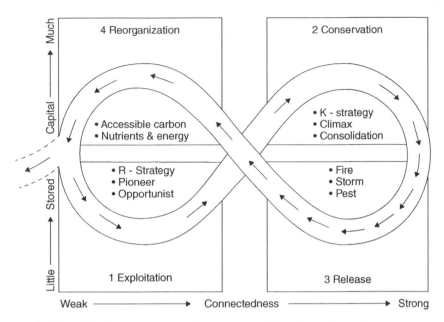

Figure 3.2   The four ecosystem functions and the flow of events between them. © Cambridge University Press 1995 (by permission). From Holling et al. (1995, 63).

Drawing from Holling, the Beijer project hangs on a dynamic and changing model of ecosystems, shown by the diagram in Figure 3.2. Ecosystems, they suggest, exist within a range of four "normal" states. First is exploitation, which involves the rapid colonization of a disturbed area, followed by conservation, a stage characterized by "slow accumulation and storage of energy" (Holling et al. 1995, 62). Holling et al. (1995) liken this first phase to the "entrepreneurial phase" of a new market, and thus followed by something more akin to the development of a "bureaucratic hierarchy" in the conservation phase. The third phase involves release, or what Holling et al. also call the "creative destruction" phase, during which "the tightly bound accumulation of biomass and nutrients" becomes more fragile until it is released. According to Hollings et al. (1995), the final phase in the ecosystem cycle is that of "reorganization," where soil processes reorganize nutrients to prepare for the next phase of exploitation.

Holling and his co-authors use economic theory to aid in explaining their ecological theory, representing the four phases as akin to that of how a firm builds and then restructures capital. They liken the "release stage" to "when an industry like IBM or General Motors accumulates rigidities to the point of crisis, followed by efforts to restructure" (62), closely following Joseph Schumpeter's concept of creative destruction.[19] The model is summarized as one where "chaos emerges from order, and order emerges from chaos!", theorizing a fast period of "release and reorganization" (chaos), and a slower period of "stability and productivity" (order). [20]

The rise of resiliency thinking has become its own cottage industry in critical social sciences.[21] For the purpose of this chapter, what is important to take from this model is that the Beijer project draws from an ecological model that is not focused on maintaining biodiversity, per se, but on maintaining *ecological resiliency*, defined as the capacity for an ecosystem to respond to disturbance by either resisting damage or recovering quickly, moving from exploitation to reorganization. If ecosystems are in states of "discontinuous change" (Perrings et al. 1992, 207), then the ecological goal, and thus the conservation goal, must not be to "*freeze* ... the diversity of the ecosystems (functional as well as biological)" (205), but rather to promote the ecological stability or resiliency that allows ecosystems to recover after disturbances. In this view, biodiversity's key ecological purpose is not its role as a "storehouse" of genetic resources or direct use benefits, but rather "its role in preserving ecosystem resilience" (205), which Perrings et al. define as an ability for ecosystems to maintain "self-organization in the face of demands for ecological services over time" (1992, 202).

Ecosystems are replete with uncertainties; for Holling et al., they are dynamic and can descend into chaos and disorganization, but it can be a productive chaos, as when a company like General Motors decides to take on new strategies like offshore production. The main concern in this ecological model is with changes that cause the system to "exit" and morph into an entirely new form, as demonstrated by the arrow at the far left hand side of the diagram in Figure 3.2, termed a "regime shift."[22] What matters is maintaining a system that provides resiliency in the provision of ecosystem services.[23]

The economics of biodiversity then, is less about diversity in and of itself; it is about finding a level of diversity that can facilitate resiliency. Unlike the Ehrlichs' rivet hypothesis, or the bioprospecting promise, this ecological-economic model – a bioeconomic promise, if you will – suggests that the line between necessary and surplus species might be found in order to bound the ideal level of conservation – not too much nor too little. As an ontology of ecosystems, resiliency refuses the notion of a singular, stable equilibrium, but simultaneously suggests that we might be able to define at least some conditions that can maintain ecosystem services. The next million dollar question is, is it possible to find "some level of biodiversity" needed for ecological resiliency and ecosystem service provision?

I will come back to Beijer after a short foray into the scientific discussions on these questions, which go back to the 1950s at least. This foray evokes the complexities involved in articulating the "unity of quality and quantity" (Mann 2007), the right measures that the economics of biodiversity demands.

How does biodiversity matter? The rivet–passenger debate

For most ecosystems and most ecosystem services, scientists lack a great deal of understanding regarding which entities of life underpin ecosystem functions. Which species and genes, and in what amounts, are needed to ensure ecosystem functions and services continue? A flip question to this is what is the effect of removing features of a system (say through extinction)? In a highly cited 1983 paper entitled "Extinction, Substitution and Ecosystem Services," Ehrlich and his Stanford biology colleague Hal Mooney demonstrate the uncertainty surrounding extinctions and ecosystem services. They contend that the effect of species extinctions on ecosystem functioning and services is highly variable: "ecosystem services can, in effect, be 'killed outright', 'nickled and dimed to death' or even enhanced by extinctions" (249). Ehrlich and Mooney suggest that the type of species extinction matters. They focus on whether the species is a controller, referring to organisms that are more crucial to the ecological structure – a species "through which the principal flows of energy and materials pass" (4). An extinction of controller species, according to Ehrlich and Mooney, has more impact on the provision of ecosystem services compared to others. However – and this is critical – identifying controllers in an ecosystem is not easy. "The degree of control exercised by a single species is often difficult to evaluate," Ehrlich and Mooney write, "even when its role is well understood" (248). The authors use this uncertainty to reaffirm the tenets of the rivet hypothesis. They cite Aldo Leopold's famous words: "To keep every cog and wheel is the first precaution of intelligent tinkering" (1949, 252). However, what the 1983 article reveals is that Leopold's dictum, and the rivet hypothesis, are *hypotheses*.

This research area is not new; early ecologists focused intensively on the relationship between species diversity and ecosystem stability. Charles Elton (1958) and others like MacArthur (1955) and Hutchinson (1959) all hypothesized that the relationship between diversity and stability was a positive one: increased diversity leads to increased ecosystem stability. In 1969, a high profile scientific symposium titled "Diversity and Stability in Ecological Systems" addressed the question head-on, but participants were unable to come to any agreement on the definition and measurement of the relationship between diversity and stability (Farnham 2007).

In the early 1990s, in the same years as the Beijer project, Hal Mooney launched and chaired a Scientific Committee on the Problems of the Environment (SCOPE) program to take up this huge question about the relationship between biodiversity and ecosystem function.[24] The debates pivot between two hypotheses about the relationship between diversity and function: rivet and passenger. The first, as we know, posits that any type of species loss reduces the structural integrity of the system (the rivets on the airplane). The second thesis, initially proposed by Brian Walker (1992), is nicknamed the "passenger" hypothesis because it proposes that "most species are superfluous, more like passengers than rivets, and that only a few key species are needed to keep the system in motion" (Baskin 1994a, 202). The passenger hypothesis suggests that the line between necessary and redundant species can and should be known in order to guide a more efficient conservation effort. Much hangs in the balance of this debate. If one adopts a strong "rivet hypothesis" approach, then there is no need for calculation because we simply must preserve everything. However, if one adopts a strong "passenger" hypothesis, calculation and modeling are needed to facilitate the ranking.

Reporting on the SCOPE project, Mooney, writing with Stuart Chapin and Ernst-Detlef Schulze (Chapin et al. 1992), sidesteps this debate by claiming that redundancy of species is important for the "insurance factor" they provide. They mean that several species of the same guild (functional type) will likely react differently to any extreme event (e.g. climatic change), and so that in a time of changing climate, redundancy strengthens ecosystem resiliency. As such, diverse species may not be necessary for ecosystem functions and services, but they are necessary on an evolutionary timescale, especially in a time of changing climates. At the same time, Chapin, Schulze, and Mooney attach themselves to the passenger hypothesis, arguing that it is important to identify what they call "keystone species," a functional group without redundancy (like the "controllers" described above) because of their more integral position in the ecosystem. As this classic "middle path" suggests, the SCOPE project did not say definitely that diversity matters or does not.[25] As science writer Yvonne Baskin summarized in her book on the SCOPE project, it did make "it clear that species are not ecologically equal; some are more valuable in terms of services rendered than others" (1998, 8). But even if this is the case, due to limitations in knowledge, she goes on to say, ecologists cannot "declare any species, however humble, expendable" (8).

Investigation into the relationships between diversity and ecosystem services continues, a key question in conservation and ecological studies, from applied science to theoretical modeling.[26] What it reveals is that this is a complex and ongoing area of ecological research that continues to scupper the production of quantified, universal value related to biodiversity. Rendering biological diversity legible for management, into entities whose worth can be adjudicated by an ecological-economic tribunal (and thus ranked and ordered) is a persistent problem. Yet, perhaps ironically, the very material-semiotic object that is so resistant to singular measures of value, biodiversity, is itself an object produced less than a decade prior by the same institutions and circuits that struggle with it in the Beijer project on biodiversity.

Getting On with the Adjudicating: What Is Necessary and What Is Surplus

In laying out their ecological model, Holling and his collaborators draw from the "passenger" hypothesis of ecosystem functioning. However, their model goes further in stating that "only a small set of species and physical processes are critical in forming the structure and overall behavior of terrestrial ecosystems" (1995, 67), such as the set of grass species that make up the productivity of savannas, or the species of birds that mediate budworm outbreaks in the boreal forest. In other words, all species are not necessary to maintain ecosystem functioning and scientists ought to work to define these boundaries between those who are critical rivets and therefore necessary, and those who are passengers and therefore redundant.

The passenger hypothesis is a way of thinking about the role of biological diversity in ecosystems that is far more compatible with the approaches and goals of economics than the goal of saving every rivet. Where the rivet hypothesis, at least in theory, treats unequals, or the dissimilar, equally (because we cannot distinguish between necessary and surplus species), the ecological-economic model out of the Beijer focuses on differentiating among the characteristics and functions of components of biological diversity. This ecological-economic model aims to align quality with quantity, to make, in the words of Frankfurt School maestros Max Horkheimer and Theodor Adorno, the "dissimilar comparable" (1944, 4–5). Among Beijer participants, the project of making the dissimilar comparable was not necessarily for the purpose of producing relative prices, or even exchange value; instead participants discussed the need to undertake a prioritization of conservation resources. One must mark in quantitative form the differential contributions of nonhuman living things in order to identify the most rational and efficient form of conservation – that elusive MS* (Figure 3.1 above). A 1994 IUCN report titled *The Economic Value of Biodiversity* and co-authored by Beijer participants captures this transformation, and situates this shifting ecological consensus as politically pragmatic, as the only way forward:

> It is in that respect "too late" for a good deal of the world's biological diversity. If so, it is essential to choose between different areas of policy intervention – not everything can be saved. This view is reinforced by the fact that the world is extremely unlikely to devote major resources to biodiversity conservation. The issue then becomes one of using the existing budgets as wisely as possible. If not everything is to be saved then a ranking procedure is required (Pearce and Moran 1994, 32).

In the realm of ecological-economic-market articulations, it is simply not efficient, or rational, or even *possible*, to value all nonhumans equally, especially in times of austerity. Or as David Harvey quips, quoting Thomas Jefferson: one of the laws of liberal capitalism is that "there is nothing more unequal than the equal treatment of unequals" (2001, 178).

Such a vision of ecological systems that can distinguish between necessary and surplus is more compatible with the notion that conservation should focus on maintaining *levels* of biological diversity, rather than the "preservation of all species" (Perrings et al. 1995c, 4). What is needed, as Charles Perrings and his co-authors write, is more work on setting the rates of biodiversity loss and the minimum threshold levels that will ensure "human welfare and existence." Reflecting the cruel optimism of the dismal science almost too perfectly, this statement that biodiversity conservation is not about "complete preservation of all species in the world" is repeated often; participants clearly saw this as an important academic and policy-guiding contribution.

Yet, while many chapters and articles emerging from the Beijer project on biodiversity write about species redundancy as a hopeful conclusion to the project, other chapters and articles (sometimes by the same authors) tempered this point. For example, ecologists Holling et al. state outright that their theories should not be read as an argument "that some species are more important in nature but most are not" (1995, 70). Rather, they claim that while "passenger species" may not be necessary to the provision of ecosystem services now, they still provide the material for future evolutionary processes; they are *insurance* species. Holling et al. use an example from the Jurassic period to underline this point: Who would have predicted, they write, "the explosive ascendancy of mammals" (71) that followed the age of the dinosaurs? In other words, all these other passenger species "provide options for such unpredictable futures" (71). Similar points are made from the summary of the SCOPE program in 1992, conversations taking place amongst ecologists.[27]

Yet this scientific ambiguity is tempered by the political-economic context, by what is seen as a pragmatic realization of the very marginality of their debates. Holling and his collaborators write later in the same paper, "By focusing exclusively on those species [insurance species] in attempts to preserve biodiversity is to imply a recipe for diluting time, energy and money" (Holling et al. 1995, 71). Even if they could not provide guidelines for adjudicating between species that are crucial and those that are not, "relentless pressures" "of agriculture, development, and population" as Baskin reports from the SCOPE meetings, are making conversations about "species triage unavoidable" (202). Not all can be saved, the line goes, and so "ecologists would like to provide guidelines to help policymakers make the best cuts" (Baskin 1994a, 202). Walker, the ecologist who developed the passenger hypothesis, argues that a policy that places "equal emphasis on every species is ecologically unsound and tactically unachievable" (quoted in Baskin 1994b, 658). Demonstrating the lamentable pragmatism that provides so much fuel to the project of enterprising nature, even Paul Ehrlich, the king of the rivet hypothesis, concedes that detailed research could help with making "optimal allocation of limited funds to conservation efforts, and the relative merits of competing parties' interests" (Ehrlich 1998, x). I, for one, have a hard time reading these statements without feeling the tragic weight of the realpolitik, in terms of both the marginal nature of ecology but also the ease with which they seem to embrace a pragmatic complicity.

Through the Beijer, biodiversity loss is turned into what the participants see as a more manageable, rational, and calculable problem: not saving all species but, rather, defining a fixed limit to biodiversity loss within which ecosystem resiliency can continue, thereby ensuring the provision of global to local ecosystem services. The very nature of the "biodiversity crisis" is altered – at least in theory; it is not a problem of saving all but some. If the rise of biodiversity is rooted in a belief that we cannot distinguish species that are necessary or surplus to ecological functioning, to planetary health, or to profit (as argued in chapter 2), the experts around the Beijer Institute suggest that such boundaries must be made knowable to guide the ecological-economic tribunal. We cannot invest in all, so we need to identify morphologies that are redundant, or surplus to human need. In a paradoxical flip, the economics of biodiversity is not only a play to shape an economically rational approach to biodiversity conservation, but also becomes a potential site of rationalizing and normalizing biodiversity loss.

The "endless accounting" of the Beijer Institute

The economics of biodiversity is a *potential* site of rationalizing and normalizing biodiversity loss. I say potential because demarking this line between necessary and surplus remains a huge challenge, to return to the diversity-function debate. At what point do ecosystems become more open to undesirable regime shifts? And what role does biological diversity play in this?[28] The Beijer is exemplary in both embracing economic rationality – we must choose, rank, and order; we cannot save all – and rejecting economic rationality – yet we still have no clue how to rank and order! And, even, for whom do we rank and order?

The tensions go further. Participating experts embraced dynamic ecological systems while at the same time seeking to render these systems into equilibrium models of supply and demand. The "economics of biodiversity" wants to hold together a nonlinear and surprise-ridden model of ecological systems but also to nail down "the *appropriate* level of exploitation – as well as conservation – of our natural environment" (Barbier et al. 1994, 18). As a whole, the "economics of biodiversity" rejects resource exploitation models focused on "maximum sustained yield," while still seeking the "sweet spot" of optimal allocation, a "sweet spot" that is recognized as enormously difficult to identify.

What can we make of these tensions and paradoxes? As Morgan Robertson (2006, 369) argues, such "inevitable paradoxes" emerge when two or more disciplines try to become legible to each other, given their somewhat autonomous logics and norms. Indeed, the "synthesis" of ecology and economics requires both rejecting equilibrium and embracing it at the same time. The clean lines and graphs they produce belie these troublesome tensions.

Yet I think these paradoxes can tell us something about the idealized modes of governance in Western modernity, about deeply rooted imaginaries and dreams. The "economics of biodiversity" reflects the ongoing force of Enlightenment knowledge production, a knowledge-seeking endeavor that is calculative and utility driven (Horkheimer and Adorno 1944), that promises to end the uncertainties and fearful futures of life on Earth. Any moment of paradox or tension – such as those put up by dynamic, surprise-ridden ecosystems – is solved via a call for further knowledge, rather than resulting in any assessment of fundamental limitations of the project. More interdisciplinary knowledge is needed, or a new model that can comprehensively bring together more information and data, especially ecological and economic (see chapter 4). Niklas Luhmann goes as far to conceive of modernity as a project of aligning multiple (and multiplying) arenas of expertise and systemic knowledge that "when sutured together, ostensibly provide panoptic knowledge" (cited in Robertson 2006, 369). In other words, a key task of modern expertise is to draw together different disciplines and knowledges so as to create a more comprehensive picture of how the world works, a panoptic, all-seeing vision. Such a project is always incomplete, but the promise of completion encourages more and more investments in expertise, and in the promise of the "grand synthesis" of knowledge and power. The word promise is thus very important.

Taking on the promise of internalized externalities, especially in the carbon market, Larry Lohmann aptly describes this as a project of "endless accounting," where the promise of equilibrium, of a perfectly functioning market – the MS* in Swanson's graph (Figure 3.1, above) – fuels "indefinite further investment in centres of calculation," requiring "contributions of ever-expanding bodies of expertise" (Lohmann 2009, 529). This point of Lohmann's, I argue, can be expanded in conversation with Horkheimer and Adorno's famed critique of the Enlightenment. For Horkheimer and Adorno, the Enlightenment promises human liberation from nature through quantification, calculation, and rationality. The Enlightenment project, for them, more than anything, optimistically finds emancipation for humanity in Western science and knowledge. But, and this is crucial for Horkheimer and Adorno, it is not just any form of science and knowledge. In the panoptic patchwork of expert knowledges, those who can turn qualitative relations, the unknowns of life on earth, into differences that can be rendered as quantity and thus commensurable are most prominent and influential. In the Enlightenment promise, it is Lohmann's "endless accounting," positivistic objectivity, that promises to make the future calculable and legible, leading to human emancipation. For Horkheimer and Adorno, the irony is that this tendency toward quantification, toward calculation of utility, is underpinned by a desire to control and dominate that is driven by a deep-seated fear of the unknown and unfamiliar – of all that does not "conform to the standard of calculability and utility" (3). The Enlightenment is driven by a desire to liberate that is based on a controlling and dominating impulse, on desire for the mastery of nature.

In the Beijer project, we see this Enlightenment promise on full display; the project is one that seeks to liberate humans from the dangers of changing ecological systems, but does so through further domination and control. This domination and control, this mastery of nature, is seen in the calls for the right measurements, the right ways to "articulate the unity of quality and quantity" (Mann 2007, 27). Such ways of representing the rich qualitative and dynamic relations of ecosystems are needed to produce an economic tribunal, a tribunal that can produce not just any difference, but comparable differences, differences that can be ranked – not just subjectively (by you or me), but differences that are supposedly neutral and objective. Difference, then, is apprehended in only one sphere, in one register – the quantitative. This is, I argue, a reason why such a tribunal, even an ecological-economic tribunal that aims to conserve biological diversity, ends up in a situation where it could work to legitimize biodiversity loss. This form of differentiation erases other ways of apprehending value; it attempts to erase qualitative difference (Mann 2007). But too, and here is where value meets Enlightenment dreams, it also aims to excise uncertainty; it aims to make the unknown known in the pursuit of human emancipation from the oppressive nature of the unpredictable future.

In the next section I draw out an example of an ecological-economic tribunal that does precisely this: it aims to guide crane conservation by transforming crane diversity into quantitative forms that can be ranked, guiding the investment of scarce resources.

An Economic Tribunal for Cranes (Group *Gruiformes*)

Economist Martin Weitzman, a participant in the Beijer, developed a diversity function to guide conservation investments between the 15 crane species (see Table 3.1 for a list), research published in one of the oldest and most prestigious journals in economics, *The Quarterly Journal of Economics*. Weitzman's tribunal is composed of several sources of data, beginning with the probability of extinction for each crane (see Table 3.1), which he characterizes as "best guesses" (1993, 161) determined in consultation with several crane experts. The second dataset is a measure of the "value of diversity" (162). Weitzman comes up with his own measure, arguing that the magnitude of loss from extinction should be measured in relation to "how different the extinct crane is from the surviving cranes" (164). If the extinct crane is closely (genetically) related to another species, then its value is less and its loss less significant than the value of a loss of a species that does not have another closely related species. While not necessarily surplus, the species is marked as "more killable," to use Donna Haraway's term. Drawing from DNA experiments conducted by biologists, Weitzman compiles a table estimating the "genetic distances" between all 15 crane species (see Table 3.2).[29] The numbers in the table refer to the mismatch in underlying DNA for each crane species – the higher the number, the higher the difference.

In many ways, this "value of diversity" measurement is a measurement of what Mooney, as discussed in chapter 2 of this book, and Holling et al. (1995) in this chapter, describe as the insurance or option value of species.[30] It is a measurement that prioritizes species conservation based on a goal of maximizing genetic variation. Weitzman recognizes that "genetic difference is not the only possible measure of diversity" and that "we could argue for a long time about whether or not the genetic distances … are the appropriate inputs to use for a diversity measure" (1993, 166). But he challenges his critics to come up with a better formulation, especially given existing

Table 3.1 Crane Extinction Probability. © Oxford University Press (by permission). From Weitzman (1993, 161).

Number	Common name	Scientific name	Geographical range	Extinction probability
1	Black crowned	*Balearica pavonina*	Central Africa	0.19
2	Grey crowned	*Balearica regulorum*	South-East Africa	0.06
3	Demoiselle	*Anthropoides virgo*	Central Asia	0.02
4	Blue	*Anthropoides paradisea*	South Africa	0.10
5	Wattled	*Bugeranus carunculatus*	South-East Africa	0.23
6	Siberian	*Grus leucogeranus*	Asia	0.35
7	Sandhill	*Grus canadensis*	North America	0.01
8	Sarus	*Grus antigone*	South-East Asia	0.05
9	Brolga	*Grus rubicunda*	Australia	0.04
10	White-naped	*Grus vipio*	East Asia	0.21
11	Eurasian	*Grus grus*	Europe, Asia	0.02
12	Hooded	*Grus monachus*	East Asia	0.17
13	Whooping	*Grus americana*	North America	0.35
14	Black-necked	*Grus nigricollis*	Himalayan Asia	0.16
15	Red-crowned	*Grus japonensis*	East Asia	0.29

ecological data. We don't have a quantified representation of crane contribution to ecosystem function or resiliency, but there is one for genetic variability (see Table 3.2).

Weitzman then combines the extinction probability of each crane species with their genetic differences to produce what is called "marginal diversity" (see Table 3.3), which is essentially a measure of the relative payoff in diversity from improving the survival prospects of each crane species. The higher the number, the greater the payoff from each unit of investment in conservation of that particular species. A very likely-to-go-extinct bird such as the whooping crane (species number 13) scores relatively low because of its genetic closeness to

Table 3.2 Genetic differences between cranes. © Oxford University Press (by permission). From Weitzman (1993, 165).

							Species number								
	1	2	3	4	5	6	7	8	9	10	11	12	13	14	15
1	0	86	417	382	392	362	384	372	393	389	336	388	399	364	390
2	86	0	382	387	408	348	392	368	362	401	355	400	371	351	360
3	417	382	0	60	113	180	141	149	123	150	110	104	142	156	147
4	382	387	60	0	138	191	137	173	109	156	111	117	138	117	168
5	392	408	113	138	0	142	116	143	138	168	115	148	129	121	103
6	362	348	180	191	142	0	140	107	125	190	121	143	166	144	144
7	384	392	141	137	116	140	0	136	143	145	114	138	151	176	138
8	372	368	149	173	143	107	136	0	54	71	112	167	138	146	120
9	393	362	123	109	138	125	143	54	0	105	111	135	154	181	124
10	389	401	150	156	168	190	145	71	105	0	145	144	180	166	129
11	336	355	110	111	115	121	114	112	111	145	0	7	29	53	24
12	388	400	104	117	148	143	138	167	135	144	7	0	43	63	33
13	399	371	142	138	129	166	151	138	154	180	29	43	0	72	62
14	364	351	156	117	121	144	176	146	181	166	53	63	72	0	59
15	390	360	147	168	103	144	138	120	124	129	24	33	62	59	0

Units. degrees centigrade multiplied by 100.
Note: Numbers 1–15 across the vertical and horizontal axes refer to crane numbers found in Table 3.1.

Table 3.3 Conservation diagnostics. © Oxford University Press (by permission). From Weitzman (1993, 172).

Species number i	Probability of extinction $P(i)$	Marginal diversity $-\dfrac{dV}{dP(i)}$	Elasticity of diversity $-\left(\dfrac{dV}{dP}(i)\right)*\left(\dfrac{P(i)}{V}\right)$
1 (Black crowned)	0.19	8.7	11.3
2 (Grey crowned)	0.06	14.1	5.8
3 (Demoiselle)	0.02	7.0	0.9
4 (Blue)	0.10	4.8	3.3
5 (Wattled)	0.23	7.8	12.3
6 (Siberian)	0.35	10.3	24.6
7 (Sandhill)	0.01	11.1	0.8
8 (Sarus)	0.05	4.7	1.6
9 (Brolga)	0.04	6.5	1.8
10 (White-naped)	0.21	9.2	13.1
11 (Eurasian)	0.02	1.3	0.2
12 (Hooded)	0.17	1.4	1.6
13 (Whooping)	0.35	4.5	10.7
14 (Black-necked)	0.16	5.8	6.3
15 (Red-crowned)	0.29	2.9	5.7
		100.	100.

other cranes. The biggest beneficial impact for diversity, according to this schema, would be gained by focusing attention on the grey crowned crane from southeast Africa (species number 2).

Obtaining "the most effective diversity-improving investment strategy from a global perspective" (1993, 173), though, would require the calculation of the relative costs of each unit of conservation investment for each species. Such a measure, if it existed, could be combined with the measure of marginal diversity to come up with a numerical ranking of which species conservation efforts would truly yield the "most bang for the buck." Without this information, Weitzman calculates what he calls the elasticity of diversity (the final column), which shows how making a species "safe" in terms of extinction risk (i.e. lowering its probability of extinction) increases the gains in diversity. The point of this measure is to show that some species are better investments than others not only because of their genetic uniqueness but also because they have a better chance of recovery. The Siberian crane is exemplary in this regard, because for any decrease in extinction probability, it will yield almost twice the expected diversity gains.

Weitzman's diversity function exemplifies the incorporation of biodiversity into the laboratory of economics, and econometrics. It accomplishes what Lohmann (2009) describes as "imposing distance": it expresses value in abstract and remote ways. The value of the crane is expressed not in aesthetic, ethical, emotional terms or even in terms of the relationships these birds might have with humans or nonhumans living near these cranes (or far away). The function is not even really concerned with the living, breathing birds, but with their genetic differences, the value they hold for genetic variability, a very narrow interpretation of Holling's ecological resiliency model. In this ecological-economic tribunal, adjudication for conservation investment is located not within the seas of subjectivity, but rather within neutral, objective accounting of the genetic make-up of the cranes themselves (which surely cannot lie!) and the probability of their extinction. Which crane, Weitzman's modeling aims to show, is more or less enterprising, which crane will deliver the most bang for investment buck over the long term, and live to tell about it?

Of course, as decades of science studies research show us, and as Haraway points out so clearly, none of this is neutral or objective. The god trick is just that, a trick. It directs our attention away from the human actors that do the work of designing models, doing science, from the cultural, institutional, and knowledge bias, the uneven relations that make it possible for some to construct apparently "valueless" ways of producing value. In Weitzman's example, too, we see the way that this tribunal wreaks (potentially) rationalized violence. Some species offer better bang for the buck, as least in terms of their genetic diversity and survivability, some are more necessary and others are surplus, or at least, "lost causes." The Eurasian crane, viewed through Weitzman's functions, is simply not worth the effort.

Rationalizing Biological Diversity Loss

This chapter began with the IUCN's 1988 call for a synthesis of economics and ecology that would quantify the "indirect values" of biodiversity to human life and existence. IUCN sought to gain support for the case that – once indirect values were accounted for – conservation was an economically rational policy choice. Theirs was a call that the Beijer Institute looked to answer as it drew together economists and ecologists in pursuit of a new knowledge synthesis, a synthesis deemed crucial for life on earth. They established a new field, the economics of biodiversity, suturing together a common understanding between ecologists and economists once at odds.[31]

This field sets for itself an enormous task. It aims to translate the dynamic and poorly understood ecological relations and processes that biological diversity underpins into equivalent measures of usefulness that can be compared and ranked, a universal measure of use value. Through new ecological-economic calculative methodologies, one might be able to choose, for example, which was the better conservation investment: a demoiselle crane of central Asia or a black-crowned crane from central Africa. This ability to differentiate living beings' universal use values would support the maintenance of ecological resilience needed to ensure human survival over the long term, the argument goes. But this adjudication of species cannot be separated from power relations; it is also about directing how to live, how to organize human–nonhuman relations, how to *live* properly. An ecological-economic tribunal is not only about rendering unruly and complex ecosystems visible and legible for governance; it is also about rendering unruly and complex socio-ecologies into their most proper and properly efficient forms.

Quantified, supposedly universal values are not neutral, then, just like the god trick; claims to a universal value from a distance exclude other cultural and political modes of knowing, valuing, living. For example, the 1994 IUCN report *The Economic Value of Biodiversity*, produced by Beijer participant David Pearce and Dominic Moran, states as fact that moral arguments must go:

> If all biological resources have "rights" presumably it is not possible to choose between the extinction of one set of them rather than another. All losses become morally wrong. But biodiversity loss continues apace because of the reasons we have cited and for one other we have not mentioned: the competition between mankind and other species for the available space. The reality is that little can be done to prevent huge increases in the world's population.

Pearce and Moran go on to conclude that these constraints mean it is already "too late" for most biological diversity. And furthermore, constrained government funds require conservationists and governments to make choices with their resources; "not everything can be saved" (32).

The question of the rights of nonhumans is a lengthy and dense one; here I want to flag that Pearce and Moran reject other ways of ascertaining value, including moral, in favor of what they see as the pragmatic approach to the problem. For them, an ecological-economic tribunal is the way forward, sidelining a massive and complex set of historical, moral, cultural, and political questions about how to live on the planet with diverse nonhuman others. Such an approach is, as Wendy Brown (2005) argues, a hallmark of neoliberal rationality. This is a rationality that "relieves the discrepancy between economic and moral behavior by configuring morality entirely as a matter of rational deliberation about costs, benefits and consequences" (42). The neoliberal citizen, she aptly says, is a Benthamite, calculating rather than rule abiding.

I am not saying that universal nonhuman rights are the path to take, or that we should return to the "known unknown" limit of biodiversity. Rather, that Pearce and Moran ask us to consider *these* particular choices, to choose between biodiversity conservation as a moral issue, a universal question of rights, and an ecological-economic tribunal, exposes a deeply rooted characteristic of modernity, a tendency to universalize problems, to search for a single grid of intelligibility. This characteristic of market society is part of what drives violence and dispossession, even when it aims to address or reduce them. As Anna Tsing (2005) explains, "Those who claim to be in touch with the universal are notoriously bad at seeing the limits and exclusions of their knowledge" (8).

The Beijer project sets out to make biodiversity loss tractable for modern governance. Yet it also holds within it the specter of further rationalizing biodiversity loss, of more god trick "saving" violences from on high. I say "more" because rationalized violences are not new to biodiversity conservation. But here we see a different sort of rationalized violence, one that yet again produces ideals about what and how to live with nonhuman others from a distance, in the service of the global body politic, all of humanity. Both the forms of violence of protected area dispossession and the potential forms of violence of an ecological-economic tribunal acutely demonstrate the biopolitical rationality at the heart of liberal governance. This is a logic that is oriented to fostering the health of citizens, with survival and optimization taking place "against a micro-grind of scarcity, supply and demand and moral value adjudication" (Brown 2005, 40). Indeed, a crucial thread that weaves between biodiversity as a "known unknown limit" demanding conservation of all and biodiversity as more flexible "level" is their adherence to the ordering force of scarcity: scarcity of resources – biophysical, monetary, and political.

Waiting for the Tribunal

Yet the economic tribunal desired by many gathered around the Beijer Institute remains just that, a desire or a will. Despite producing a disciplinary apparatus that aligns so well with prevailing neoliberal governing ideologies and practices (this is the early 1990s), international conservation action and planning does not become guided by Weitzman-like equations showing which species are rivets and which are passengers. The very lack of economic logic in conservation continues to be an object of criticism in the conservation world, up to the present. For example, writing a decade later, and echoing points raised in the

Beijer project on biodiversity, one Nature Conservancy staff member writes that conservation investments have not paid enough attention to "serious cost-benefit analysis" which is "essential for the strategic allocation of scarce conservation resources" (Cleary 2006, 733). Another article in *Biological Conservation* laments the fact that while conservation organizations seek efficiency in their resources, "none explicitly include costs" (Murdoch et al. 2007, 376); the article proposes the adoption of a formal "return on investment" framework.[32]

Articulating "the unity of quality and quantity" (Mann 2007, 27) that could make biodiversity legible and tractable in modern governance is a tricky business indeed. Writing in 2010, over 15 years after the Beijer project, project lead Charles Perrings notes:

> Not surprisingly, the specification and estimation of ecological-economic production functions that capture both the jointness of the production of ecosystem services, the interactions between services, and the impact of changes in abundance of species is still in its infancy (5).

What this jargon-filled quote says is that greasing the wheels for a desired ecological-economic tribunal continues to flummox the most committed economist; it is still difficult to measure biological diversity and what it does. And so, the ecological-economic tribunal remains elusive, more an ideal than a program of work.

Here we are, once again, waiting for the fulfillment of bioeconomic promises, awaiting the right representations and knowledge that can unleash the utopian dreams not necessarily of profit (as with bioprospecting), but of equilibrium, an equilibrium that can tell us the right, most efficient, and truthful relations between humans and nonhumans. Chapter 4 looks at the rise of ecosystem service science and modeling, a scientific shift that aims to produce the kinds of measures needed for an ecological-economic tribunal.

Notes

1   Enthusiasm for bioprospecting as incentive for conservation in the tropics peaked in 1991 when pharmaceutical giant Merck signed a 10-year, $1.3 million deal with the Costa Rican National Biodiversity Institute (INBio). But, InBio notwithstanding, bioprospecting has largely failed to deliver on its promises of both profits and conservation (Castree 2003a, Mateo et al. 2001, Firn 2003, Burtis 2008, Dalton 2004, Clapp and Crook 2002). This is in part because the rates of discovery of viable biochemical compounds for medical research are extremely low, with the vast majority of natural

products found in plants and microbes unlikely to contain the potent biological activity needed for pharmaceutical use (Firn 2003, 212). In drug development from natural products, only between 1 in 10 000 and 1 in 40 000 compounds screened is likely to yield a marketable product, and of those compounds that do reach clinical trials, fewer than 1 in 4 will be approved as a new drug (Clapp and Cook 2002). Others suggest that the conservation goals and poor business acumen of environmentalists further harm the economic viability of bioprospecting (Clapp and Crook 2002, Burtis 2008). A dedication to benefit sharing with local Indigenous popula-tions, for example, is cited as a significant contributor to Shaman Pharmaceu-ticals' bankruptcy in 2001 (Clapp and Crook 2002). In a similar vein, Dalton (2004) suggests that the push for benefit-sharing agreements by Indigenous populations has de-incentivized participation by large pharmaceutical inves-tors (see also Burtis 2008). Others cite bureaucratic barriers in the post-Rio context as disincentives to profitable bioprospecting (Dalton 2004).

2   Only recently (2014) did the "third objective" of the Convention on Biological Diversity pass into a stronger international form with the "Nagoya Protocol." This Protocol is a fleshed out international regulatory framework governing the "access and benefit sharing" from sovereign genetic resources.

3   In this way the Beijer Institute is a kind of "forum of articulation," a space where disciplines become understandable and legible to each other, where they develop ways of speaking to each other and creating new alignments of knowledge and power (see Luhmann 1989, 2002; Robertson 2006). Forums of articulation are important spaces because modern society, as Morgan Robertson writes (drawing from Luhmann), can be understood as a patchwork of "multiple, specialized knowledge systems, each with their own governing logic and standards of verity" (2006, 369). These knowledge systems, Robertson writes, must at times take in information or input from other "outside knowledge systems" requiring some form of translation.

4   A major contribution of political ecological research and theorizing is show-ing the marginalizations and violences of metropole-produced ecological crises (Forsyth 2002, Fairhead and Leach 1995).

5   Biodiversity loss was also chosen (over all other environmental issues) as an attempt to counter the dominance of climate change in the scientific and policy agenda, a dominance that suggests, for Perrings et al. (1995a, 1) that (1) the "social cost of biodiversity loss is trivial compared to the cost of climate change" or (2) the return on investment in conservation is "much lower" than the return on investment in climate change prevention. The project of Beijer participants, then, was motivated also by the goal of creating new economic and ecological knowledge and evidence that could convince governments to invest more in conservation. If only we could create the knowledge necessary for biodiversity to be always and forevermore included in governance processes, the participants seemed to be saying.

6   For a full list of participants see Perrings (1995b, xv).

7   These were (1) the role of biodiversity in keeping ecosystems resilient in the face of external disturbances; (2) the value of biodiversity in generating

ecological services; (3) the social and economic forces driving the loss of biodiversity; and (4) the measures required to reduce or even reverse the current rate of biodiversity loss (Barbier et al. 1994, xii).

8   See, for example, Turner et al. (1995), Brown and Roughgarden (1995), Perrings and Walker (1995).

9   The subjects and approaches found within these books and journal articles are diverse. For example, articles focus on valuing the contributions of marine parks (Dixon et al. 1994), traditional knowledge, biodiversity, and resilience (Berkes et al. 1995), applications of Computable General Equilibrium models to deforestation in Costa Rica (Burgess 1995), rangeland ecology (Walker 1994), coastal and estuarine ecosystems (Costanza et al. 1995), and international regulation of biodiversity decline (Swanson 1995).

10   Elsewhere, Beijer participant Buzz Holling explicitly rejects an approach to understanding biological systems as "discrete elements in space and time, predictable in isolation from other elements in the ecosystem" (Holling et al. 1998, 347), which are the telltale markings of resource sciences like forestry and fisheries. See also Beijer participants Joan Roughgarden and her co-author Fraser Smith (1996) and economists David Pearce and Charles Perrings (1994), making similar points.

11   See the three books (Barbier et al. 1994; Perrings et al. 1995b, 1995c) and journal articles (e.g. Perrings and Opschoor 1994).

12   Swanson is an important and recurrent figure in global biodiversity politics from the outset. In the 1980s, he was a strong advocate for property rights for genetic resources to improve incentives for conservation, and, more recently, he has been working on proposals to create a Green Development Mechanism for biodiversity, akin to the Clean Development Mechanism – the subject of chapter 7.

13   Swanson spends a good part of his chapter explaining the graph; here I provide some more context. The vertical axis is both the benefits and costs. The graph depicts a scenario with little chance of halting conversions. This is demonstrated by the "Marginal benefit of conversions" downward *curving* line – where each unit of land converted actually reduces costs due to increasingly specialized technological processes or inputs like machinery, chemicals, and improved agricultural species. This means the benefit of each unit of land converted (the area above the curve) is increasing at each point in time because the land converted is becoming more and more profitable (rather than the opposite). Swanson argues that this conversion scenario is deepened because there is a slow-moving demand-related constraint, represented by the linear downward-sloping "Demand for conversions" line, what Swanson terms a "demand curve" (230). For Swanson, the demand for conversion declines as consumers become hesitant to accept substitutes for products produced by converted lands (say they prefer wild meat over beef). The rate (and slope) of that decline in benefit, Swanson argues, depends upon how willing consumers are to accept the substitutions. If consumers are willing to accept substitutes, the line could have a less steep demand constraint and thus result in

complete conversion of land (as in the top "Demand for conversions" straight line). If they do not, the slope of the demand curve is more rapid, leading to a halting of conversions. This line is also related, Swanson argues, to population growth, which causes the demand curve to be more like the top straight line sloping downwards (i.e. more benefits to conversion). As this sounds, responsibility for a potential "global conversion process" that decimates the "natural evolutionary process" is located within the decisions of consumers.

14   Challenges led Beijer director Charles Perrings, along with Carl Folke and Karl-Goran Maler, to declare: "A major task for future research is to fix the boundaries for sustainable levels of biodiversity and ecological services with greater precision than has been possible in the past, and to explicitly state the time and space scales" (1992, 202).

15   Proposed by Theodore Panayoutou and supported by the Beijer (see Perrings and Opschoor 1994), this system draws its inspiration from systems to preserve historic buildings in urban areas. In the context of national or regional conservation, a tradable development permit system would involve zoning two broad categories of lands: those that are open for development (less valuable for conservation) and those that are important for biodiversity conservation. Individuals who own land in the conservation area would sell their permits to those in the development area for a price that would cover the costs of conserving that patch of land, covering the "the net present value of the income stream from the forgone development opportunity" (Panayoutou 1995, 310). A similar system is proposed for use in the international sphere, where the potential buyers (in a voluntary system) could be "environmental organizations, foundations, corporations, governments, pharmaceutical companies, scientific society, universities, etc" (cited in Barbier et al. 1994, 203). As I explain in chapter 7, such notions of tradable credits become part of the contemporary global biodiversity policy debate in 2010.

16   This notion is drawn from a well-known 1952 paper by Ciriacy-Wantrup.

17   In addition to these debates over setting limits and the order of policies, Beijer participants also debated whether it is possible, preferable, and/or necessary to set such limits on a macro or global level. This relates to debates about the limits of economic growth more broadly, such as the approach to steady-state economics advocated by ecological economist Herman Daly (1973, 1977), in which material throughput is severely restricted. However, some Beijer-participating economists like Pearce and Perrings rule out the idea of such standards at the global level.

18   Recognizing the range of policy approaches, Perrings and Opschoor write that there is no reason to choose between "command and control" and "market instruments" – both approaches can be used to limit externalities by "address[ing] the gap between the private and the social cost of resource use ... both are designed to improve efficiency in the allocation of resources by confronting users with the true cost of their actions" (1994, 5).

19 Holling et al. (1995, 62) write: "This last phase is essentially equivalent to processes of innovation and restructuring in an industry or in a society – the kind of economic processes and policies that come to practical attention at times of economic recession or social transformation."

20 As a bit of an aside, there are similarities between the models advanced by Holling et al. and some notions of self-organization in economics. For example, economist Kenneth Arrow notes that economic systems are not completely prone to single equilibrium, despite his earlier prominent role in defining the core neoclassical concept of general equilibrium. He argues: "It is clear that many empirical phenomena are not covered well by either the theoretical or empirical analyses based on linear stochastic systems, sometimes not by either. The presence and persistence of cyclical fluctuations in the economy as a whole of irregular timing and amplitude are not consistent with a view that an economy tends to return to equilibrium states after any disturbance" (cited in Foster 1993, 975). Arrow is saying that economies work in ways we cannot always predict, and they evolve in new directions after major disruptions.

21 See, for example, Walker and Cooper (2011), Turner (2014).

22 A good example of such a shift is the pine beetle outbreak in western Canada, where shifting climate and monoculture forestry bred conditions for the rapid diffusion of the beetle, which had infested 723 cubic meters of timber as of 2013; the Province of British Columbia has estimated that 57% of the pine volume in the province may be destroyed by 2021 (Province of BC 2013). Such a shift is being described as a regime shift (Raffa et al. 2008, Starzomski 2013), where the ecosystem leaves the figure eight loop into a "qualitatively different state" (Holling et al. 1995, 66), one that – at least from the point of view of the British Columbian government and communities that depend on these forests – is less productive.

23 In general, as noted above, the concept describes the "services" that ecosystems provide, such as "watershed protection, climate stabilization, erosion control, etc" (Barbier et al. 1994, 17) and "on which economic activity and human welfare depend." Concerns with ecological resiliency and system shifts are to do more than anything with the possibility of vast shifts in the provision of ecosystem services, which would compromise economic activity and human welfare. The concept of ecosystem services is critical for suturing together ecology and economics; its concern with the functioning of ecological systems entwines with the way they provide "service" for human communities. In this way, the concept operates as a boundary concept between the two disciplines.

24 Formed by the International Council of Scientific Unions (now called International Council for Science) and based in Paris, SCOPE is a nongovernmental body with 38 member nations, and access to what are considered "the world's best scientists" (Mooney and Lubchenco 1997, xiv). The program on biodiversity and ecosystem function drew together hundreds of scientists around the world to address this relationship between diversity and ecosystem function and stability. A participant in the project, Ian

Noble (an Australian ecologist), nicely characterized the problem as one of "lacked evidence." "Ecologists," he argued, "have convinced policymakers of the importance of biodiversity to ecosystem health" and "we now find ourselves having to produce the evidence" (cited in Baskin 1998, 202).

25   The project did produce some general conclusions, noting that diversity mattered little for some functions like nutrient cycling and decomposition, and it mattered only up to a point in terms of ecosystem productivity (see Baskin 1994a, 202). Reflecting on the project, one soil scientist put it frankly, "It's impossible to say we need every beetle to maintain function" (quoted in Baskin 1994a, 202).

26   It is hard to dabble in this area as there is so much literature. Some recent articles include Thompson and Starzomski (2007), Cusens et al. (2012), Biggs et al. (2012), Isbell et al. (2011), Bullock et al. (2011), Rey Benayas et al. (2009), Ridder (2008).

27   See, for example, Chapin et al. (1992).

28   This question is raised in the analysis of the functional role of wild grasses found in Barbier et al. (1994), a chapter in one of the Beijer books. For feeding livestock, domesticated grasses can substitute well for (or sometimes better than) wild native grasses. If we are only considering the feeding of cattle (single commodity production), Barbier et al. (1994, 16) argue, "then there may be little additional value to be gained from maintaining the diversity of grasses in rangelands." However, if the goal is broadened to include the maintenance of ecosystem functioning, then both types of grasses maybe necessary. The scientific question that must be answered is this: At what level of wild grass decline does the "resiliency" of the ecosystem break down?

29   DNA hybridization research is conducted predominantly to assess evolutionary relationships between species.

30   Holling et al. argue that focusing on insurance species is "a recipe for diluting time, energy and money" (1995, 71), but here Weitzman identifies *the* insurance species, the most genetically distinct and the most-likely-to-survive crane species.

31   Writing in the foreword to one of the Beijer book collections, economist Partha Dasgupta states: "It was a privilege to be allowed to attend the meetings at which these studies were planned and then presented. It was also a delight, because it was at these meetings that I saw how folk who spoke different languages gradually learned to understand one another and found that they could not only do business, but that the particular business of the day was both inspiring and pleasurable" (Dasgupta 1995, ix).

32   Naidoo et al. (2006) argue that without estimates of cost, NGO or conservationist claims of wise investment and efficient allocations of effort remain speculative.

4

Ecosystem Services as Political-Scientific Strategy

In a 2009 interview, Peter Kareiva, the chief scientist at The Nature Conservancy (TNC), one of the world's largest conservation organizations, spoke frankly about his embrace of ecosystem services (ES): "Biodiversity is one thing that some people, like point zero zero zero zero one percent of the world's population, care about." Unlike biodiversity, however, "there are all these other ecosystem services that people care about that are some dimension of nature." Responding with surprise over our peanuts and beer, I noted that Kareiva should probably look for a new job, given how central biodiversity is to TNC programming. He laughed. "Yeah."

Until 2015, Kareiva not only remained VP and chief scientist at TNC; he also became a key "bomb thrower" (Voosen 2011) in a heated debate about the future of conservation. Ideas like Kareiva's are at the forefront of a debate over what is being called the "new conservation."[1] Whereas the "old" conservation was focused on conserving and protecting biodiversity hotspots or so-called "pristine" nature, the "new conservation" rejects Edenic baselines (Robbins 2014). For example, Kareiva (with co-authors Michelle Marvier and Robert Lalasz) writes that conservation must stop focusing on a return to "pristine, prehuman landscapes"; it must "jettison ... idealized notions of nature, parks, and wilderness ... and forge a more optimistic, human-friendly vision" (Kareiva et al. 2012, n.p.). If the past is no longer the guide for conservation, then what might replace it? The new conservation foregrounds human needs as the most crucial measure in the calculus of how to intervene. It is oriented toward understanding the services that ecologies provide – toward ecosystem services.

Enterprising Nature: Economics, Markets, and Finance in Global Biodiversity Politics, First Edition. Jessica Dempsey.
© 2016 John Wiley & Sons, Ltd. Published 2016 by John Wiley & Sons, Ltd.

As Kareiva explained to me in his charismatic, plain-speaking manner, ecosystem service science is all about "making it clear to people that the consequences of stupid ecosystem decisions come back to haunt them." This, I argue, is the new conservation's ethos in a nutshell. Avoiding "stupid decisions" is a goal that is hard to dispute. No one wants to make stupid decisions: not businesses, not governments, not individuals, not communities, not families. But how do we figure out how to make *good* decisions related to ecosystems and biodiversity, ones that we can be sure will be good not only next year, but in 20 or even 50 years? This is perhaps the million-dollar question of ecosystem science, policy, and environmentalism over the past several decades, as experts struggle with the unknowns of massive atmospheric and ecological changes, changes now described with the shorthand designation "the Anthropocene." The new conservation re-enacts a major Enlightenment theme: the best decisions will come from the right application of reason, especially positivistic science.

Yet, right application of reason rarely serves all, often benefiting some bodies over others, some communities over others. Along these lines, the ecosystem service turn has come under the critical lens of academics, especially critical geographers, anthropologists, and political ecologists. An ecosystem services approach, critical scholars (including myself) argue, risks reducing complex ecosystems to market logic, laying the ground for new rounds of accumulation and profiteering, even, potentially, for rationalizing biodiversity loss (as I discussed in chapter 3).[2] Sian Sullivan (2010a), for example, describes the rise of ecosystem services as another instance of an "imperial ecology," given that ecosystem services – as metaphors, scientific practices, models, and policy approaches – are largely formulated by experts located in institutions of global North, and largely aim to serve "transcendental corporate capital and finance" (119). The ecosystem services apparatus is economistic and Western-centric, yes, with the potential for all kinds of injustices. Yet it is also an "internally conflicted and polyvalent project" (Dempsey and Robertson 2012, 759) – a project involving a wide range of actors with diverse viewpoints on science, policy, and politics.

And so, what can we learn about ecosystem services by starting in the middle of things? In this chapter, the "middle of things" means talking to some of the most ardent supporters of ecosystem services, like Peter Karieva, Walter Reid, and Gretchen Daily, in order to understand their logics and rationales.[3] It also means studying a technoscientific innovation, a calculative device known as InVEST (Integrated Valuation of Ecosystem Services and Trade-offs tool) that aims to operationalize the ecological-economic tribunal described in the chapter 3. In some ways, the actors and their institutions I explore exemplify the "imperial nature" of the ecosystem service project promoted by American and

European-based institutions like The Nature Conservancy, World Resources Institute (WRI) and Fauna & Flora International. The circuits of power and knowledge of nature conservation are by no means global; they are well-traveled routes etched out over decades, now deep grooves resistant to change. Yet those circuits are not the whole story in the rise of ecosystem services. Neither can broadbrush gestures toward the structuring effects of capitalist or colonial logics offer complete understanding. And so, to the middle of things.

The Rise of Ecosystem Services

My first encounter with the concept of ecosystem services was during a round of negotiations of the Convention on Biological Diversity (CBD) in 2006, where I was editing the daily NGO dossier the *ECO*. As Table 4.1 shows, ecosystem services arrived suddenly into the Convention, entering the text of decisions quickly around the mid-2000s. In response to the increased attention on ecosystem services in the decisions of the CBD, the editorial of the *ECO* rang out: "Biodiversity provides benefits – it is a public asset, not a service!" "Like a naïve slip of tongue or a bad fad," the text exclaimed, "the language of the World Bank seems to have gripped hold of CBD documents" (ECO 2005, 1), referring to the explosion of ES terminology and the "monetary valuation of ecosystems" in the texts under consideration.[4]

The explanation for this rapid rise of this language in the decisions is simple: the 192 government signatories to the CBD had begun considering the results of the titanic Millennium Ecosystem Assessment (MA). Conducted between 2001 and 2005, the MA assessed the conditions and trends of the world's ecosystems through an ecosystem service frame-

Table 4.1   Growth in ecosystem services discourse at the CBD.[5]

Conference of the Parties	# of decisions mentioning "ecosystem services"	# of times words "ecosystem services" found in decisions
COP 5 (2000)	2	5
COP 6 (2002)	3	3
COP 7 (2004)	6	15
COP 8 (2006)	7	79
COP 9 (2008)	10	52
COP 10 (2010)	22	118
COP 11 (2012)	16	83
COP 12 (2014)	13	57

work. The assessment involved over 1300 experts worldwide and was funded by the United Nations Environment Programme (UNEP), the Global Environmental Facility, and several private foundations and governments. Headed by Walter Reid, the MA's objective was to provide an appraisal of the condition and trends in the world's ecosystems, the services they provide, and the options to restore, conserve, or enhance the sustainable use of ecosystems.[6]

The ecosystem services concept has a longer history, of course – it was not invented by the Millennium Ecosystem Assessment. In their 1981 book *Extinction*, the Ehrlichs were writing about the services of species well before the idea of biodiversity was in wide circulation (chapter 2).[7] It is a concept also enmeshed in the development of ecological economics as a heterodox branch of economics in the 1970s, and following on this, is a crucial concept in the Beijer Institute discussions in the early 90s.[8] In 1997, a banner year for the concept, Gretchen Daily released her book *Nature's Services*, and Robert Costanza valued all of nature's services for the first time.[9] So while Table 4.1 shows a meteoric rise of ES in CBD language in 2006, the term has at least a 30-year history in academic circles.

For global biodiversity conservation, the embrace of ES-related ideas in the mid-2000s was a departure. Throughout the 1990s and 2000s, the practice of global biodiversity conservation by NGOs, biologists, and ecologists had focused on the establishment of parks and protected areas, and, in the development world, on "integrated conservation and development projects" – projects that emphasized joint objectives of environmental protection and economic development. Very little within these approaches was justified through quantitative assessments of the ecosystem services provided; conservation organizations focused on protecting as many species and unique spaces as possible.[10] Conservation criteria were based upon assessments of rarity and threat or aimed to achieve representative conservation of all the major "ecoregions" on earth (what WWF defines as distinct assemblages of species, natural communities, and environmental conditions). Most conservation organizations did not define their priorities based on the contributions of a particular ecosystem, tract of land, or assemblage of species to provide ecosystem services, or according to price or return on investment.[11]

So what propelled an academic concept to the forefront of international negotiations and conservation discourse more broadly? To answer this, I start with Walter Reid, whom I interviewed in 2009. Several interviewees said that the MA was the brainchild of Reid, who told me that the idea for this massive assessment developed through conversations he had with other scientists, economists, and conservationists in and around the WRI, where he was the vice president for science between 1987 and 1998. During his early days at WRI, Reid, in collaboration with scientists from IUCN and WWF, was integral to establishing biodiversity as a global entity of concern through the creation of the CBD. By the end of his tenure he had turned his attention toward ecosystem services and an assessment that could appraise their state globally.

Millennium Ecosystem Assessment as geopolitical environmental strategy

Reid explained that, during the first half of his tenure at WRI, he "was not thinking about ecosystem services per se, and definitely not using the language." However, by the latter half of Reid's tenure at WRI, the discourse began to shift. Reid noted that it was "somewhere about halfway through that ten year period I distinctly remember spending a bunch more time talking about ecosystem services with people, and thinking about it." This period corresponded loosely with the Beijer Institute's project on the economics of biodiversity (see chapter 3), but also a broader shift in environmental governance toward, at least in some spheres, market-oriented policies.[12]

Reid mentioned his conversations with Jane Lubchenco, who was a Pew Scholar with Daily and Costanza and also a frequent participant in Beijer events. At some point, Reid explained, Lubchenco began telling him "that WRI should do more on ecosystem services." He said: "At the time I didn't see how. I saw it was a valuable concept, I got the concept ... but I couldn't quite see where we could get political traction." For Reid, ecosystem services seemed "more valuable from a scientific, conceptual standpoint but in terms of policy [he] couldn't see where you went." Reid was hesitant to believe that the concept would perform well in politics.

According to Reid, the idea for the MA arose in a brainstorming session at WRI in preparation for the World Resources Report for 2000, a report being prepared in partnership with the United Nations Development Programme (UNDP), the United Nations Environment Programme (UNEP), and the World Bank. In the course of discussions, participants decided that the millennial report could lead to a wider assessment, something more comprehensive, along the lines of the Intergovernmental Panel on Climate Change (IPCC) reports. The idea of the assessment quickly received assent from top representatives of these key international institutions (UNDP, UNEP, and the World Bank).

Reid said that MA proponents saw the initiative as being geopolitically strategic. Past attempts to conduct international biodiversity assessments had failed miserably due to ongoing resistance from government leaders in the Global South. For example, the first Global Biodiversity

Assessment (GBA), conducted by Hal Mooney in the mid-1990s, could not be endorsed by the Parties to the CBD because of concerns of the Southern governments. The GBA, Reid explained: "was a complete failure predominately because of the concern of developing countries that this was something that was not in their interests. An international assessment of biodiversity would start pointing fingers at tropical countries because they are losing biodiversity, so why on earth would they want to be part of it?" For Reid, an ES framework could reach across the North–South divide and could more effectively address development and equity issues:

> With biodiversity you immediately think about threatened species and extinction, but with ecosystem services – all of a sudden you position it as a development issue, the benefits countries can get from these systems. So the way I would talk with people if I talked in that framing, the ability to quickly make sense to a developing country, of why they might be interested in this was quite simple and straightforward. And it didn't lead immediately to the split in natural and modified systems, you can talk about the ecosystem services provided by agricultural systems. It was a much more effective way to have international assessment be positively received by the developing countries, which were the biggest hurdle we faced.

For Reid, the ecosystem services concept was politically effective because it "mapped into the economic and development interests of the country." He concluded, "It was a political motivation that drew us to that framing, but it was [for] economic reasons that political motivation worked and so that is what gave it that staying power." This distinction is important. In chapter 2, I described how biodiversity conservation was positioned as being in the national, developmental state interest because of the promissory exchange values in genetic resources. But Reid points out the limitations of a biodiversity focus, which, despite the potential values from bioprospecting, was not easily represented as being in the best interest of Southern governments. Ecosystem services, on the other hand, for Reid, could gain traction in the international realm, could bridge the North–South divide because it is a more overtly *economic* concept, linked to the development interests of a country.

Reid went on to head the MA, assembling a star cast of ecologists, including Mooney and World Bank Chief Scientist Robert Watson, as well as economists like Partha Dasgupta, an eminent scholar based at Cambridge. The MA was, like the Beijer biodiversity project, a site where ecologists and economists could work together. Reid explained that while ecology and economics shared some conceptual similarities, there was still "not a whole lot of space for them to work together until you started talking about ecosystem services." And in building the team, Reid said, MA staffers purposely sought out mainstream economists: "We went out of our way to identify the lead economists for the working group as people who would be perceived by neoclassical economists as one of them ... We went out of our way to get blue chip, mainstream economists, because of the concern that we did not want [the assessment] to be readily dismissed." (The concern that mainstream economists would dismiss ecosystem services comes out of a debate that occurred following Costanza's 1997 estimate of the value of the earth's ecosystem services.)[13]

The MA did not undertake any new science, but rather focused on synthesizing previous ecosystem knowledge within the new framework of ecosystem services. It developed a formal definition of the concept that now appears frequently on PowerPoint slides, policy documents, and in the academic literature, highlighting the four categories of benefits people obtain from ecosystems – provisioning, regulating, cultural, and supporting services (see Figure 4.1 below). As the diagram demonstrates, biodiversity is the larger "box" or superstructure from which ecosystem services flow. In its results, the MA charted a grim decline in ecosystem service provision, estimating that 60% (15 out of 24) of the ecosystem services examined were in the process of being degraded or used unsustainably. Importantly, the MA also argued that the effects of ecosystem service degradation were felt most acutely by the poorest peoples, claims that would not surprise geographers who study ongoing enclosures and processes of uneven development.

Whereas previous global biodiversity assessments focused on assessing changes in numbers of species, the MA focused on changes in the capacity of ecosystems to provide services. A crucial distinction from what came before was a focus not so much on the "impacts" or a necessarily negative human encroachment into "pristine nature" and more emphasis on the kind of trade-offs involved when deciding, say, between turning a mangrove forest into a shrimp farm (a provisioning service) or leaving it as is to provide climate and flood regulation.

Ecosystem Services in Global Conservation: Failure Is Everywhere

Geopolitical strategy was one – but not the only – rationale for these experts' embrace of ES. Many of the scholars, experts, and leaders I interviewed explained that conservation's past failures were the primary reason for their turn to this new concept. Despite the expansion of global

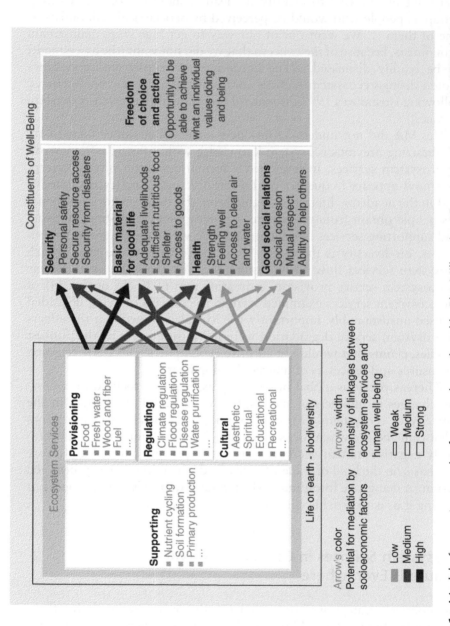

Figure 4.1 Model of ecosystem service framework employed by the Millennium Ecosystem Assessment. © World Resources Institute 2005 (by Creative Commons license). From *Millennium Ecosystem Assessment* (2005, vi).

biodiversity law and policy and of endangered species legislation, and despite the growth in protected areas, everyone I interviewed described global biodiversity conservation as a failed project. For interviewees, ES provide a corrective framework for conservation. Writing in *Conservation Biology*, heavy hitters in the ecological and conservation biology world (including Daily, Ehrlich, and Kareiva) proclaimed:

> Conservation biology began life as a crisis discipline, its central tenet to understand and help reverse losses of biodiversity and habitat. Those losses continue unabated, implying that, as a discipline, *we are failing in our central charge* ... Nature for Nature's sake resonates only with the already converted. Business interests, farmers, and the billion humans living in rural poverty remain unwilling or unable to move. We need these people as partners in conservation, and ecosystem-service approaches provide a means of motivating and enabling them. (Armsworth et al. 2007, 1383; italics added)

Once again, this band of conservation scientists and advocates claim that conservation failures exist because not enough people care about nature for its own sake. The discipline and practice of conservation, they argue, must frame ecological changes in direct relation to humans in order to resonate with more people – i.e. with business, farmers, and the "billions living in rural poverty."

An emphasis on ecosystem services is not only politically pragmatic, Armsworth et al. claim; for these authors, a focus on the benefits humans gain from ecosystems also provides much-needed disruption of an "outdated and dangerous" conservation perspective, one that places "*Homo sapiens* as somehow detached and insulated from ecosystem processes" (Armsworth et al. 2007, 1384). This latter quote would not sound out of place in William Cronon's (1995) famed essay, "The Trouble with Wilderness." In that essay, Cronon critiques the idea, history, and practice of wilderness conservation (and much biodiversity conservation) as one that keeps the "human ... entirely outside the natural" (80) and offers little help in confronting the environmental problems we face.

While Armsworth and his colleagues do not elaborate on the dangers stemming from conservation that places humans outside of nature, they clearly want to challenge what they call an outdated idea of human separateness from nature. For Armsworth et al., an approach that emphasizes the benefits to people provided by ecosystems provides the best way to enlist the support of diverse groups of people – including the rural poor – in the conservation of biodiversity. Their work thus follows previous scholarship arguing that the idea of human separateness from nature has, historically, resulted in unjust conservation decisions.[14] A focus on ES, for Armsworth et al., provides the basis of

more socially inclusive conservation, one that can focus on socioecological relationships. This view departs from a kind of misanthropy in
conservation, as another university-based ecologist explained to me in
an interview:

> Ecologists were studying quote–unquote pristine ecosystems, and they
> would study human impact on those systems. And any impact they could see
> from humans was automatically put into the bad category. From a policy
> perspective it did not lead anywhere. It was always people bad, get rid of
> people. If people disappeared from the planet, the planet would be better.

This focus on embedding humans within conservation science and practice
is another theme – along with the failures of past conservation efforts – that
emerged when I asked interviewees about the rationale for ES. As the same
university-based ecologist explained, "if you implement an ecosystem
service framework for decision-making, you automatically start asking
who benefits." Or, for another university-based ecologist: "It's the only
case where I've seen people make explicit, in conservation decisions,
anything about human impacts." She continued, "Because you at least
think about the people before you kick them out of somewhere and you at
least have to pay attention to how they make their living, and what their
condition is, and how it's going to change based on your decision."

Ecosystem services and economic valuation: Weighing trade-offs

This attempted disruption of nature–culture dualisms is only one strand in
ES's rise, of course. The ecosystem services concept brings together a wide
range of rationales. For many conservationists and ecologists, the failure
of biodiversity conservation is due not only to the exclusion of human
needs or the violences of eviction, but also to the fact that conservation has
not engaged enough with economics, and especially, with economic value.
Heather Tallis, a scientist with the Stanford-based Natural Capital Project,
and Peter Kareiva explain it clearly in one of their many articles on this
subject: "Without economic valuation, decision-makers and governments
implicitly assign ecosystem processes a value of zero, and not surprisingly,
then select actions that reap rewards according to values everyone already
understands (like a factory)" (Tallis and Kareiva 2006, 747). As this observation suggests, the root forces of ecological change, and biodiversity loss,
are – as another interviewee said – "all economic at heart."

"Look," explained one of my interviewees, this time an ecological
economist based at an environmental organization, "we ... live in a
capitalist system, dictated by market economies and market exchange."

The benefit of ES approaches, she said, is that they allow decision-makers to see the value of ecosystems, to "appreciate their scarcity and use them more responsibly." The argument is that by focusing on changes in ecosystem services and, ideally, economically valuing them, "agents" (i.e. governments, consumers, businesses) will be able to rationally weigh the options and rank a more comprehensive set of "trade-offs" between different courses of action or developments, rather than considering only the revenue and jobs that an economic activity might produce. In assessing and valuing ecosystem services, the goal is – as Kareiva had it – to avoid "stupid ecosystem decisions" that come back to "bite" governments, decisions that, for him and others, fail to adequately consider the full range of ecological and economic trade-offs.

While an ES approach seeks economic valuation, interviewees suggested, the idea was not necessarily to commodify nature or create markets. In my interview with him, Walter Reid emphasized the limited nature of market-making opportunities in ES – opportunities, he said, that were isolated to perhaps forest carbon and maybe a few water funds. Many interviewees registered similar reservations, a fact that might signal a departure from the celebratory age of the bioeconomy found in the conservation discourse of the late 1980s and early 1990s, a discourse focused on the potential of genetic resources and bioprospecting (see chapter 2). One ecological-economist working at a US university said that he didn't think that the "commodification of nature" was "what the ecosystem services concept was about." He continued: "I mean, to me, it's about recognizing that you know nature has value to people whether there's a market or not." Highlighting economic values through an ES approach, he said, might lead to the implementation of any number of policy instruments, and he reiterated that the approach aims to "sharpen what the trade-offs are" and "hopefully allow us to make better decisions."

Overall, for many interviewees, the economic valuation of ecosystems within an ES approach was, first and foremost, a way to "open doors." An ecosystem service framing could multiply the "receptive audience ten-fold or a hundred-fold," as Walter Reid told me. Or as Peter Kareiva explained: "Our marketing team hadn't heard of ecosystem services … but they discovered biodiversity didn't work" in terms of communication. In contrast, he said, the ES concept resonates with governments, resource agencies, and the business community. "Biodiversity closes doors," this interviewee said, summarizing. "Ecosystem services opens doors." Of course, in part, ecosystem services are a more strategic rhetorical approach than biodiversity, simply because of the concept's focus on human well-being. However, interviewees reiterated that the possibility of economic valuation, or the expression of the value of services in terms of the supposedly universal equivalent of money, is crucial to "opening doors." As one academic ecologist noted: "Even just that simple reframing and the ability to talk about it sometimes in dollar terms gets people to the table much more quickly than anything I've ever done before ... It allows us to put information that is usually really intangible into terms that are immediately recognizable to decision-makers." Propelling the rise of ecosystem services, I argue, is the desire to make a nature that capital or the state can see (Robertson 2006), but also a nature that anyone and everyone can see. The economic valuation of nature as part of ES is about creating a kind of commensurability that all people – from bureaucrats to finance ministers to farmers to you and me – can recognize.

Ecosystem Services as Political-Scientific Strategy

Ecosystem services should be understood as a kind of political-scientific strategy, an attempt to meaningfully intervene in twenty-first-century politics.[15] Promoters of an ecosystem-service approach seek to change political and economic systems so these systems recognize the underappre-ciated and undervalued ecological structures necessary for life on earth – this is the political. These advocates also want to better understand the impacts of changing ecologies – this is the science. Together, as a political-scientific strategy, an ES approach seeks to produce new *interests*.

Interests? Ecosystem service experts aim to produce ecosystems as entities in which someone has a stake. A state, for example, should be interested not only in new hydropower dams for electricity generation to fuel their economic development, but also in the forests that regulate water flow needed for the efficient operation of these dams. A coffee farmer or a multinational investor in coffee should be interested not only in the price of coffee, but also in the forests that foster wild pollinators that can improve crop yields. The management of interests, argues Foucault (2008) in his analysis of liberal governance, is a principle to which governmental reason, and thus also political strategy, must conform.[16] Liberal governance is less concerned with things in them-selves, like individuals, land, or wealth, and more so with how "things," like land or say biological diversity, *interest* individuals or collectives.

As a political-scientific strategy, ecosystem services are a renewed attempt to bring ecosystems to the political table through the production of an interest – a state, firm, community, or personal interest – in it. There is, of course, no guarantee that any interest will be created within a subject (individual or institution) or that, if it is, it will last, or succeed in the cacophony and indeterminacy of the political world. But when we see ecosystem services as a political-scientific strategy, I argue, we can better understand the logic behind it. As I described above, critical scholars have ascribed significant power to the ES turn and criticized it as a reflection of capitalist and especially neoliberal logic. But an ecosystem services approach does not always aim to set the stage for new commodities; many advocates emphatically deny that ES opens up a new site for accumulation, often understanding their science as an attempt to intervene in and regulate status quo economic development. Furthermore, when we understand ES as a political-scientific strategy we are reminded that the scientists involved are not corporate puppets, but actors with motivations, sophisticated knowledge, and epistemological frameworks who cannot be reduced to mere representatives of the neoliberal context. Viewing ES as an "interest-producing machine" can help us to recognize the multiple rationalities and logics at play. Such openness can help us to identify real opportunities for critical intervention or even alliances with scientists and experts.

At the same time, however, the ES approach is rife with epistemological problems and political risks. The very concept of an "interest," to begin with, is one that emerges out of a particular Western historical context and is tied up with notions of individualism and autonomy that are central to the Enlightenment project and capitalist social relations (Mann 2007). That people can or should relate to nature through "interests" – or that an interest is the form of relation that should be politically or socially recognized – is a very narrow way to conceive of humans' diverse connections with the nonhuman world. And as with all technoscientific objects, ecosystem services is not simply a benign technology, it is productive – it has real effects on socioecological relations, and these effects carry risks. In particular, ES could reinscribe status quo relations or, worse, deepen the inequities that mark contemporary politics.

One risk is that new ecosystem functions, especially the reproductive capacities of ecosystems, will in fact be privatized and turned into commodities – even if scientists and conservationists do not have this intention. This is a point made by many activists I worked with around the CBD, who were concerned about the rise of ES language in the Convention. "Life is not for sale" is a mantra of global justice activists concerned with the specter of ecosystem service markets. Critical scholars also make this point; Morgan Robertson (2012) is perhaps most forceful in stating that we are living in an epochal moment where "ecosystem services" are making a new social world. He compares the shift to that of individual human labors becoming social labor under capitalism, arguing that an ES turn creates abstract social nature that allows for new forms of accumulation, new ways to make money not off nature's goods, but off its services. There is no doubt that new kinds of commodities have been created to deal with environmental problems, like wetland offsets and mangrove forest carbon offsets, although as I show in chapters 6 and 7, there is reason to be skeptical about the future of these markets. There is also reason to avoid a kind of determinism when it comes to the new ecological-economic object of ecosystem services, a concept that does not itself do things on its own, but is rather, like all technoscientific objects, produced out of "a historical system depending upon structured relations among people" (Haraway 1991, 165). Still, there is a risk of ecosystem services becoming gateways to further or new forms of commodification and enclosures, especially when it is a concept that aims to draw out in quantitative form nature's enterprising nature.

A second related risk lies with claims that ES are a "better," more accurate and thus universal representation for socioecological relations. For example, Kareiva argued that people in developing countries were more likely to connect with the idea of ecosystem services than of biodiversity:

> One of the first things I had to do at TNC was to teach a course on conservation in China. And in China they did not relate to biodiversity. But they related to ecosystem services. You go to Latin America, they care about nature – but biodiversity? They will even use the word, because they have been told to, but they really mean nature. ... When you go internationally and you talk about ecosystem services, and explain it, they say that's what we've always wanted to do. Whereas they won't say that for biodiversity.

In some ways this quote parallels the concerns of scholars critical of the concept of biodiversity as a Western concept steeped in the notions of a pristine nature separate from humans (e.g. Guyer and Richards 1996). However, the idea that a different (but still wholly Western) representation of nature–society relationships – ecosystem services – could be "the universal" representation we have all been waiting for is also deeply problematic; it perpetuates the colonial tendency to prescribe how others ought to live from yet another Western location of, as Sullivan writes, once again imposing the "unidimensionalizing value system of the west" (2010a, 121).[17] In chapter 3, I described problems with the attempt to produce an ecological-economic tribunal that could adjudicate among nonhuman lives via the supposed neutrality of accounting. With ES, these risks continue: the approach seeks to embed economic calculus into the wider social fabric of value articulation, in pursuit of perfect allocation or equilibrium through the application of science. The ecosystem-service turn can be seen as an attempt to produce universal, putatively value-free value that can make interests in ecosystems equivalent and comparable, thus transforming deeply political questions about how to live into ones animated by a belief in "moral value-neutrality" of costs and benefits (Brown 2005, 40). The claim to the universal frame is where many critical scholars get itchy, particularly when this frame is produced and imposed by positivist experts far removed from the bloody, gendered, raced, and geographical socioecological histories that are a part of biodiversity loss. This is where my romance with the science becomes tepid: when it appears to be a neutral play to achieve optimization, perfect allocation, or equilibrium through the application of ecological-economic positivism. A key question is how this particular metaphor/model can avoid the colonizing, violent, and reductionist tendencies that have hampered other environmental metaphors and models. Or, if it can at all.

With full awareness of these risks, I nonetheless believe we should not dismiss ES science and policy. Rather, in viewing ES as a strategy, we retain some space for radical applications of the science, and also keep open the possibility of conversation and alliances that are needed between critical scholars and activists and the scientists and experts who develop and promote an ecosystem-service approach. To engage in this dialogue, I argue, we need to understand – with some detail – the opportunities and challenges within the specific kinds of models ecosystem service advocates are creating. In the next section, I take readers into a different middle of things, turning my attention to one model, a "calculative device": the program called InVEST. This section is detailed; this is so you – reader – can understand the tools and explore the risks and opportunities along with me.

Calculating Trade-offs: InVEST as a Calculative Device

The Integrated Valuation of Ecosystem Services and Trade-offs (InVEST) tool is an open-source computer-modeling software tool that quantifies, maps, and values ecosystem services. InVEST aims to make quantitative the *trade-offs* associated with different courses of action.[18] It attempts to render the future of flow of ecosystem services from any patch of land legible – and sometimes economically legible – to inform science-based, present-day decision-making. These models fit within a longer lineage of modeling human–nonhuman interactions through mathematical equations and abstractions (Robbins 2013), as well as the history of simulation models that aim to turn uncertain futures into quantitative predictions or probabilities that can direct policy action (Hacking 1990).

InVEST is a product of well-traveled circuits in the enterprising nature project. It is a creation of the Natural Capital Project ("NatCap"), a partnership between Stanford University's Woods Institute for the

Environment, the University of Minnesota's Institute on the Environment, The Nature Conservancy, and the World Wildlife Fund. Directors include Gretchen Daily and Peter Kareiva, and strategic advisors include Paul Ehrlich, Hal Mooney, and Walter Reid.[19] The Natural Capital Project self-describes as a collaboration focused on "developing tools for quantifying the values of natural capital in clear, credible, and practical ways" (Natural Capital Project 2011); InVEST is one such tool.

Calculating trade-offs, producing interests

Central to the InVEST tool is the idea of trade-offs. This idea relates very clearly back to the Millennium Ecosystem Assessment, which not only charted the declines in biodiversity but also – in a clear departure from previous global assessments – showed which ecological services were increasing and contributing to human well-being. For example, the MA showed humans are benefiting from increased "services" of food production and fiber production, but expansion of these services has impacts on other supporting services like water flow regulation. The impounding of streams for hydroelectric power may have negative consequences on fish populations, and thus on "provisioning services": Which do we choose? That some services are found to be increasing while others are decreasing leads us to the central governance problematic of ES science (and I would add, that of the "new conservation" as a whole): the problem of trade-offs.

The trade-offs are complex, especially if, as the MA proposes, you consider them temporally and spatially (see Figure 4.2). For example, a trade-off between maximizing a service like food production *now* in a manner that will impact the service provision *later*, or maximizing a service such as hydroelectricity in one particular place (upstream) in a manner that will impact the provision of fish in a different place (downstream). MA authors also classify trade-offs in terms of their potential reversibility, with reversibility expressing "the likelihood that the perturbed ES [ecosystem] may return to its original state if the perturbation ceases" (Rodríguez et al. 2006, 29).

With this wide range of services over space and time to consider, the number of trade-offs is virtually limitless and seemingly impossible to model comprehensively. Figure 4.2 illustrates these different categories of ecosystem trade-offs, modeling spatial and temporal scales and reversibility. The result is a demand for an almost limitless level of knowledge from a wide variety of disciplines, a kind of panoptic, universal knowledge (see Robertson 2006, 369, drawing from Luhmann 2002). In other words, ecological decision-making comes to require the development of

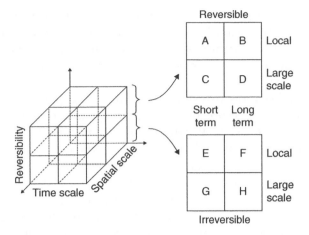

Figure 4.2 Eight categories of ecosystem service trade-offs. © Millennium Ecosystem Assessment (by permission of Island Press, Washington, DC). From *Ecosystems and Human Well-being: Scenarios by the Millennium Ecosystem Assessment* (2005, 434).

an architecture – not a physical one like Jeremy Bentham's Panopticon, but a knowledge architecture that can integrate and synthesize a huge array of knowledge about human and ecological systems, from a variety of disciplines, into one framework and model.[20]

Quantifying and modeling ecosystem dynamics is always difficult (see chapter 3); researchers find it is difficult to quantify the full impact of a single forestry operation, never mind to model and predict the changes in multiple ecosystem services in relation to one other, as well as disaggregate these services across scales and to different recipients of these services. As one lead scientist for the MA, Stephen Carpenter, wrote just after the assessment's release, "Even in the few cases where research has explored options to maximize individual services (such as crop production), there is limited research into trade-offs with other ecosystem services (such as water resources or biodiversity)" (Carpenter et al. 2006, 257).[21] And InVEST, as well as other tools, aim to fill this gap, producing not only credible research but also tools for calculating trade-offs.

InVEST, then, is a calculative device: "calculative" because it aims to create distinctions "between things or states of the world" (Callon and Muniesa 2005, 1231), not only in the present, but into the future; "device" because calculation requires the production of the models and methods that can allow users to compare different courses of action. The tool creators seek not a static estimation of services but, rather, a prediction of changes into the future. "What we look at is realistic scenarios," Gretchen Daily told me about InVEST. "We are always asking very practical, policy-oriented questions. What if we were taken this way rather than that way? It is a scenario-driven process in which we try to integrate the value of the place in its many different dimensions before and after the decision – a decision to do with resources or climate change or population change."

In sum, InVEST tools are centered on predicting quantitative biophysical differences in the future, measuring ecological-economic magnitudes between those differences. If you choose expansion of oil palm today, they tell government officials in Indonesia, then you can expect this particular land cover and ecological assemblage in 20 years, which will provide X tonnes of carbon sequestration, and Y dollars worth of flood protection, and the estimated impact on the GDP is predicted to be Z. Such quantities can then be compared to the amounts predicted to flow if government officials choose, say, the conservation of tropical forests.

The "scenario-driven process"

The InVEST model works by comparing and contrasting the amount of ecosystem services delivered in different land, water, or marine use scenarios. Scenarios describe what the future might look like given a present course of action; they require forecasting into the future based on a set of "rules" or characteristics. In analysis of a place that the modelers call the "Heart of Borneo," a region undergoing rapid land use change from extraction and oil palm, WWF and NatCap produced two scenarios: "Business as Usual" (BAU) and "Green Economy" (GE) (In what follows, much is drawn from the report of the study by Van Paddenburg et al. 2012.) These scenarios informed the production of maps of projected land use and land cover in 20 or even 50 years – maps of the future. To create this information for Borneo, the study team employed an integrated spatial software package (IDRISI Land Change Modeler) that draws from evidence of previous land use changes and the scenario "rules" to predict how the future might appear under these two scenarios. For example, the BAU scenario meant few land use planning restrictions along with active mining and expansion of oil palm plantations. The GE scenario, meanwhile, followed different rules, including the enforcement of land use plans, tax law reform, and new legislation and environmental standards. The result of this 20-year projection in Borneo is further oil palm and mining expansion in the region, presented in the two maps in Figure 4.3 – the top map is the business as usual results, the bottom the green economy. The BAU scenario results in a predicted loss of 3.2 million hectares of primary and secondary forest cover by 2020; the GE scenario only loses 0.1 million hectares (Van Paddenburg et al. 2012).

While this sounds straightforward, each step of this modeling exercise requires enormous amounts of work and data gathering.

These maps of future land use cover are only the first step. Whereas previous conservation analyses might highlight, for example, the species at risk, the InVEST software tool produces other interests in this transformation. InVEST takes these land use and land cover data sets (or marine use sets) and quantifies ecosystem services delivered between the different scenarios.[22] According to NatCap, InVEST can use data on changes in land use, land cover, and marine use to model changes in 17 ecosystem services, including blue carbon (carbon storage provided

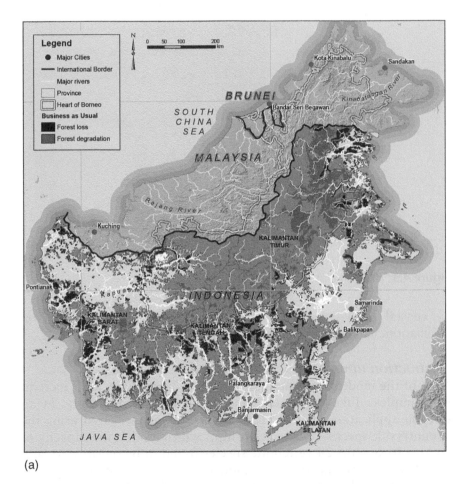

(a)

Figure 4.3 (a) Forest cover in Borneo in Business as Usual scenario; (b) Forest cover in Borneo in Green Economy scenario. © WWF HoB Global Initiative (by permission). From Van Paddenburg et al. (2012, 86).

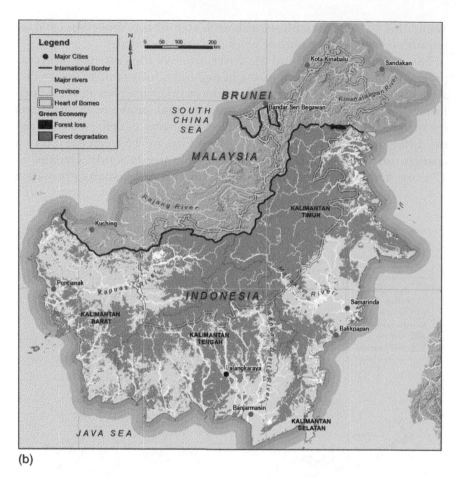

(b)

Figure 4.3 *(Continued)*

by coastal ecosystems), terrestrial carbon, coastal protection, crop pollination, habitat quality, timber production, and so on.[23]

Abstraction upon abstraction, the future emerges
Each of the models integrated into InVEST, however, is itself incredibly complex. Take the modeling of changes in biodiversity. In an Oregon application of InVEST, biodiversity was measured using the "countryside species–area relationship," which models the capacity of an area to support 24 vertebrate species found to be sensitive to land use changes. This measurement modeled the *actual* amount of suitable habitat under different scenarios against the *potential* habitat – a species' geographical range within ideal conditions. Each of the 24 species was assigned a score based on the ratio of actual habitat area

to the potential, raised to the power of "z." The "z" is user defined; lower values of z imply fewer penalties for losing small portions of habitat, and larger penalties for losing the last few units. These scores are averaged out to calculate an aggregate "biodiversity score" for each land use scenario. Researchers also calculated the "marginal biodiversity value" for each area – a measurement of the "value of habitat in the area of all species under consideration, relative to the composite value of habitat available to all species across the whole landscape" (Nelson et al. 2009, 7). They then calculated the *relative* marginal biodiversity value by measuring the change in an area's value over time. These calculations involve multiple complicated models, algorithms, abstractions, and assumptions – this analysis of the change in biodiversity in relation to future land use scenarios is incredibly complex, involving a number of steps to transform flesh-and-blood life forms and their entwined relations into single numbers, based on different land use covers.

And this is just *one* input into the model.

These layers of models result in a series of numbers and graphs that represent predicted changes of ecosystem services in each scenario. Table 4.2, for example, shows two of the results in the calculations comparing the BAU and GE scenarios in the Borneo case. Where data is available and possible, InVEST puts economic values to these flows of ecosystem services (see the "value of carbon" column in Table 4.2). For example, nutrient export is 12% higher in the BAU scenario due to oil palm production and increased fertilizer use. Relatedly, the cost of removing nitrogen from water (due to increased nutrient export in the BAU model) is estimated at US$1.9 million/year (Van Paddenburg et al. 2012). To understand

Table 4.2 Some of InVEST's calculations comparing two scenarios in the Heart of Borneo. Derived from Van Paddenburg et al. (2012, 90).

	CO_2 sequestration by 2020 (billion tonnes)	Value of carbon (billion \$US; based on a price of \$2/ tonne)	Soil retention per hectare (tonnes per hectare)	Soil retention overall (tonnes)
Business As Usual (BAU)	22.6	44.6	X	x
Green Economy (GE)	23.8	47	$x+27$ (± 12)	$x+900\,000$

each of the calculations behind these tidy numbers would take an additional book or two, so complex and detailed is each step.

In a next step, all of this forecasting is brought together to predict
economic trade-offs between the different scenarios, to give a big-picture
look at different courses of action. Modelers aggregate this information
from the Borneo case into yet another model (a System Dynamics
macroeconomic model) to show that the GE scenario produces GDP
growth that is as fast as that under the BAU (see Figure 4.4 below).
Using standard GDP accounting, they estimate that the BAU and GE
scenarios will diverge in 2020, when the environmental costs of
development will start to impact on the revenues from natural capital. In
other words, 2020 is the year when the "stupid decisions" could come
back to haunt the Heart of Borneo. In the GE scenario, "green infrastructure investments" will, for example, reduce the costs needed to
maintain hydrological services (the sediment retention, reduced floods,
improved water quality from reduced nutrient loads, etc.). These investments will also allow actors to reap ongoing revenue from non-timber
forest products, ecotourism, and biodiversity.

This is the kind of ecological-economic fact that InVEST produces, a
fact to produce interests in forests, to change the minds of decision-
makers, to show us that nature is, indeed, enterprising – if not immediately,
then at least over the longer term.

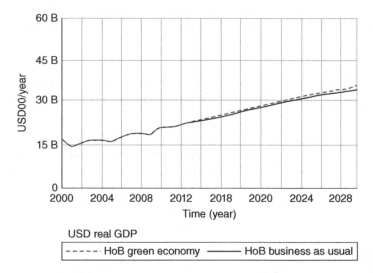

Figure 4.4 Projected Gross Domestic Product under Green Economy and
Business as Usual scenarios. © WWF HoB Global Initiative (by permission).
From Van Paddenburg et al. (2012, 85).

Ecological-economic fortune-telling

If you are feeling skeptical of these numbers, you are not alone. InVEST reports and papers are filled with caveats, notations of assumptions; no one is hiding the fact that their ecological-economic fortune-telling is filled with abstractions and exclusions. One InVEST research paper focuses on the challenges (or impossibilities) one faces in projecting future economic values or rates of growth given the enormous technological and market-related changes likely to take place in the next 20 or 50 years (Johnson et al. 2012). Kevin Anderson, from the Tyndall Center, aptly characterizes a similar problem in relation to economic forecasting for carbon offsets, hilariously noting that these models presume "powers of prediction that could have foreseen the internet and low-cost airlines following from Marconi's 1901 telegraph and the Wright Brothers' 1903 maiden flight" (cited in Lohmann 2014, 172). While these projections emerge from scientific models vetted in the world's best journals, projecting GDP changes 20 years into the future is a far cry from the positivism promoted by the likes of Karl Popper, who would cringe at the heroic assumptions and leaps of faith required to believe the predictions.[24]

Ecological dynamics are also uncertain, of course, and there are questions about how ecological thresholds or boundaries might be incorporated into ecosystem valuation. There are widely recognized limitations of ecological valuation techniques in conditions that are increasingly non-linear with climate change (TEEB 2010c). Some restoration ecologists and other scientists now write of "novel ecosystems," referring to the vast changes in the functioning of ecosystems due to rapidly changing conditions of climate and land use (Hobbs et al. 2009); such novelties make historical analogs less relevant in charting futures. InVEST model creators themselves note the lack of integration with climate models, models that predict major changes in rainfall, temperature, and other climatic variables that will also impact the delivery of ecosystem services. Yet the model still produces a stable ontology of the future upon which calculations can be based; it delivers clear, quantitative assessments of services, despite all the uncertainties and unknowns.

One significant unknown is the relationship between biodiversity and ecosystem services. The question of exactly how diversity matters for ecosystem functioning and thus services remains as it was in the early 1990s: an ongoing debate in the ecological literature (see chapter 3). Each ecologist and conservationist I interviewed raised questions about this relationship and the enormous unknowns. One ecologist (working at a major conservation organization) told me outright, "We don't know what diversity does," going on to tell me that economists "sometimes get frustrated because they just want to be told by the ecologists what's going on here"; they want ecologists to "give a probability of persistence of this service under these conditions," but he insisted, "it's super difficult to try and do that." In practical terms, this difficulty means that the InVEST model created by the Natural Capital team cannot take these relationships into account: the data simply does not exist in generalizable forms – there is no universal "biodiversity-ecosystem service function" to plunk into the model.

Ecologists face a paradox. Many who reject simplicity, who embrace stochasticity and non-linearity, must then exclude these uncertainties in their models in order to be relevant and influential. And as it turns out, InVEST itself might be too complex to be useful in decision-making. "No matter how much interdisciplinary scientists think they are over-simplifying biophysical or socio-economic processes," reads a recent NatCap paper, "decision-makers typically ask for simpler, easy-to-use and understandable decision support tools that can be readily incorporated into science-policy processes" (Ruckelshaus et al. 2013, 7).

> In the first years of NatCap, we discussed creating a tiered modelling approach with the simplest Tier 1 models providing mostly annual average outputs for single services, with no consideration of interactions among services or feedbacks. We started at this level of complexity and imagined later developing Tier 2 models that reflected daily time steps, allowed multi-service interaction and feedbacks and better reflected the complex interactions of socioecological systems. Although we have implemented some of these complexities in our suite of service models, our experience engaging in real-world decision contexts has shifted our focus to the development of even simpler "Tier 0" models that produce relative ranking outputs, to meet demand from decision-makers. (Ruckelshaus et al. 2013, 18)

What this quote tells us is that making ecosystems relevant to modern institutions of rule – from governments to firms – requires further and further simplification of complex ecological relations.

As scholars of probability and risk like Ian Hacking and Niklas Luhmann show, probabilistic thinking – the rendering of future unknowns into some imperfect quantitative assessment – plays a crucial role in contemporary governance. Though the future may be "radically contingent and unknowable," the problem that we moderns and our institutions face today is "how to act" (Reith 2004, 393). Risk calculation and quantification of future unknowns provide us with the "right way to act," even if the predictive modeling is based upon abstractions upon abstractions, radically simplified relations that extract out uncertainties. A key problem with this form of probabilistic modeling (and the mode of governance they serve) is that they excel at diffusing responsibility. If one assessment fails or has an unintended effect, decision-makers can always say, as Luhmann writes, I "decided correctly, namely in a risk-rational manner" (1993, 72); I followed the model. Thus scientific prediction of this sort is a powerful god trick that can tell us what to do from a position that is simultaneously everywhere and nowhere at once, telling the truth while evading responsibility at the same time.

But I can hear Peter Kareiva and others associated with the Natural Capital Project and InVEST in my ear: our models are rigorous, they are peer reviewed – InVEST is the Cadillac of ecosystem service models! And even a reductive prediction, they would say, is better than no science at all. It is hard to disagree with this, and as Haraway herself stated years ago, we do need to find ways "to talk about reality with more confidence than the Christian Right when they discuss the second coming and their being raptured out of the final destruction of the world" (1988, 577).

But there is a persistent and ongoing tension in the Natural Capital approach that InVEST exemplifies: it wants to change decisions so they are less stupid without changing the framework within which decisions are made. What this results in, over and over again, and in the project of enterprising nature especially, is that ecologists want to make politics more rational, science based and attentive to complexities and unknowns so as to help societies make "better decisions," but the institutions they aim to influence – the state and firms especially (as we will see in the chapter 5) – demand simple knowledge and sound bites; they demand a kind of science that seems to be quite un-scientific.

In short, as decision-makers demand ever-simpler tools and experts seek to provide them, the whole process is overshadowed by larger questions regarding the radical contingency of the future and the political accountability of actors involved, including scientists and model-makers who remain largely outsiders, despite their attempts to make their work relevant to governments.

InVEST and ruling nature

InVEST is a tool for creating maps and numbers that can produce new interests in something like land use change in Borneo. In this way, it is a tool for creating political momentum, for creating further allies in support of a conservation-oriented (or what modelers call "green economy") transition. As such, InVEST can be seen as part of a political-scientific strategy. Above all – as read from its case studies and interviews with tool creators – InVEST seeks to encourage "green infrastructure" investments. As Kareiva told me, NatCap and ES are not about market making, which he characterized as a "fad." Rather, he said, InVEST and other tools are most relevant for governments: "It's not like there are going to be these big private markets. But governments build roads, governments build dams, governments build a lot of things. So why shouldn't governments invest in ecosystem services? The natural infrastructure rather than the built infrastructure." The tool is not separate from new market making, however, as the economic case for a "green scenario" often hinges on a payment, for carbon sequestration, for example. Furthermore, "green infrastructures" have always propped up capitalist value (e.g. Mies 1998, Fraser 2014).

InVEST, then, might be better thought of as a calculative device that aims, above all, to improve capitalism, to improve market society. Making capitalism work means not only internalizing externalities, but also investing in "green infrastructure" and improving state planning and zoning. InVEST and ecosystem service science more broadly, then, are creating interests that might counter the contemporary trends of deregulation of environmental laws and policies and incentivize increased environmental monitoring and enforcement and perhaps the creation of new policies (maybe market-oriented or financial in nature, or maybe Keynesian, or even authoritarian). In this attempt to create interests, InVEST aims to draw political momentum away from the "economic-development-above-all" mantra of our neoliberal era.

At the same time, however, InVEST is a tool of depoliticization. Drawing on the work of Tania Li and James Ferguson, I see InVEST as an attempt to "render technical" the problems we face. To render a problem as in need of technical solutions – like better forecasting of eco-logical-economic change – is almost always to simultaneously render it as "non-political" (Li 2007, 7), excluding all problems that cannot be solved through a technocratic or scientific frame. This is why James Ferguson (1994) talks about the development apparatus as an "anti-politics machine": it takes very complex histories and geographies that lead to poverty and underdevelopment and transforms them into technical problems that can be solved via an eight-point plan like Jeffrey Sachs's Millennium Villages project.

InVEST and NatCap are exemplars of anti-politics, although the target of their interventions is not the development subject – the poor person in need of capital or loans, perhaps, or a mosquito net. Rather, these initiatives aim to fix the problem of vision in market societies, in capitalist states. Fingers crossed, they correct vision, hoping they will create new interests in conservation. As one InVEST co-creator told me, "I think if I understand it all right, we're just actually trying to make capitalism work the way it's meant to," a project that may involve "inter-nalizing externalities" in relatively few cases, but most definitely requires further protection and enhancement of "green infrastructures." InVEST contributes to this project by calculating the most "rational" present policy options, those producing the biggest return on investment not only now, but also into the future. By rendering ecological futures into numbers, into flows of services and economic value, InVEST seeks to identify the right intervention in the present – an intervention that is right, good, and perhaps even *just* because it will maximize value across various interests, albeit value defined in narrow, economistic terms.

An ES approach is not "bad," but rather is a project filled with tensions; it aims to bring nonhumans and ecosystems to the political table, but does so through universalizing political-scientific technologies of rule – quantification, probability, modeling, and mapping – that simplify and exclude. It is a political-scientific strategy to translate crucial ecological science into forms that can create new interests in nature, although it still remains on the margins of dominant institutions.

Ecosystem services and the quest for better decisions

Can't we avoid making stupid, shortsighted decisions? In my conversations with many ES experts, I was struck by the apparent simplicity of that demand. If we can't care for nature on "its own terms" – for each rivet – can't we at least bring it into our accounting? These sentiments are the kinds guiding the creation of the MA, NatCap, and InVEST. The basic strategy is to develop the frameworks and calculative devices that will allow decision-makers to account for a wide range of ecosystem services as interests, to make nature *live* in modern life, in our institutions. The hope is that those decision-makers will then do what is rational and efficient – to avoid "stupid decisions" that will "come back to bite us."

But is it working? Will it work? By asking these questions I am not trying to understand whether ecosystem services can bring down capitalism or end world poverty, but rather understand what the political-scientific strategy does on its own terms. Do the practical realities of ecosystem services live up to their proponents' aims? That is, do they create interests that can lead to green infrastructure investments, much-needed ecological monitoring, improved spatial planning and zoning, or stronger laws and policies? Can what seems like an anti-politics machine like InVEST chip away at bad decisions? Or is it destined to the litany of technocratic go-nowheres?

There are in fact signs that InVEST is making some inroads, with case studies in Belize, Canada, Colombia, Hawaii, Indonesia, Tanzania, and several Latin American countries. In Belize, the InVEST scenario mapping helped shape the Coastal Zone Management Plan, which the government may implement (Ruckelshaus et al. 2013). In Colombia, InVEST planning led to the reshaping of incentives for a water Payments for Ecosystem Services (PES) program toward the best sites to improve water flows (Ruckelshaus et al. 2013).

China, however, is the country wherein the NatCap seems to be having the most success creating "interests" (according to its reports). In China, InVEST is being used to quantify and assess trade-offs amongst ecosystem services. The resulting information is being used to zone "Ecosystem Function Conservation Areas," which have restricted development due to the high levels of ecosystem services they provide (see Natural Capital Project n.d.). As one interviewee excitedly noted, China now has "entire provinces designated as conservation provinces, or development provinces, and they're … basically targeting to development based on that planning and then within every province they have another plan that … basically says where activities can be done and it's all driven by biodiversity and ecosystems services." What this ecologist describes is some approximation of the vision expressed in the previous chapter by the Beijer program – a rationalized landscape managed for maximum utility of the wide variety of ecosystem services. Apparently the Chinese Academy of Sciences and Ministry of Environment has now provided training in the InVEST tool to over 200 people spanning 18 key state laboratories (Ruckelshaus et al. 2013). That the political scientific strategy of InVEST might be most effective in a decidedly illiberal state is worth noting – there are certainly fewer "interests" to manage and negotiate among.

In a self-assessment of their work in the journal *Ecological Economics*, however, the NatCap team discusses the challenges alongside their successes. They admit that they purposely chose field test sites where the chances of success and replication were high – these sites had, in their words, characteristics that include "strong leadership and partners; clearly defined authorities or decision-making pathways; and demonstrated interest in using ecosystem service information in decisions" (Ruckelshaus et al. 2013, 2). Furthermore, they recognize that some of their most intensive modeling work (such as in Borneo) has not yet yielded much change in decision-making or economic development patterns. In Borneo and Sumatra, the politics strongly favor extractive industries and monoculture agriculture like oil palm, and the "green economy" alternative projected by InVEST has not gained much political traction.

So while an ES approach is supposed to be more legible to state and capital, creating interests in conservation where none existed, ES does not yet have the widespread effect that its proponents would like. As the Natural Capital Project reflects on its own work: "The promise that BES [Biodiversity and Ecosystem Service] assessments will change policy, management, or practice for public or private sector enterprises is not yet proven." Despite "individual triumphs, the pace at which the theory of ecosystem service valuation is being incorporated into real decisions has been painstakingly slow, with disappointingly few success stories" (Ruckelshaus et al. 2013, 2). In other words, the attempt to create interests that will be taken up in governance has so far proved a halting, partial, and limited project.

Ecosystem Services: Hegemonic and Peripheral?

Some observers might say that it's too soon to provide an assessment of the political or ecological effects of the ecosystem service turn. What I suggest, however, is that we understand ecosystem services both as a dominant and problematic play for universal value and – at the same time – as a highly marginal attempt to affect flows of power and capital. That an ecosystem service approach remains largely on the margins of mainstream political economic decision-making suggests that it is doggone hard to produce new interests. Even if you use dominant economic rationalities and techniques, even if you want to make only a slight alteration to political technologies and rationalities of rule, to capitalism-as-we-know-it, the challenges are titanic. A 2013 study tallied up the globe's total "unpriced natural capital" – ecological materials and services that businesses currently do not pay for, like clean water and a stable atmosphere – and found that exactly none of the world's biggest businesses would be profitable if it had to pay for that capital (Trucost 2013). The continuing presence of the so-called stupid decisions, then, can in part be attributed to one big problem: the profit and power structure in the global political economy pivots on externalizing costs, and any attempt to change these will come up against deeply embedded power structures. Despite the rise of ES discourse in the policies of international law and in conservation organizations, and despite the creation of sophisticated modeling tools, those promoting an ecosystem service approach have struggled to operationalize it, to create real interests where none existed before. In spite of the efforts of many well-connected, now economically oriented scientists (whose models seem to fit so well with mainstream political economic institutions and logics), much of what ecosystem services offer as a political-scientific strategy toward market and societal reform is unrealized, unimplemented, peripheral. The fact that a political-scientific strategy like ecosystem services remains marginal suggests limits to the so-called pragmatic approach.

Yet I'm not convinced that ecosystem service science and models should be dismissed; they have the potential to identify the distribution of the effects of ecological change – of who benefits and who might be made more vulnerable. An ES approach might in this way help to call uneven power relations to account. In order to pose this kind of challenge, however, the ES political-scientific strategy needs to be re-thought. A strategy with a better chance of success, I suggest, is one that is a little less obsessed with trying to convince mainstream decision-makers or institutions to alter their field of visibility while maintaining status quo institutions and power relations (changing vision with the same eyeballs intact, let's say). An alternative ecosystem service could also involve ecologists and modelers building relations of technoscientific solidarity with movements who are already defending their ways of living, defending their diverse epistemologies and ontologies from the monoculturing forces that reduce their lives and lands into tidy numbers – into returns on investment, GDP, or employment rates. This is line with what James Ferguson (1994, 286) argues at the end of *The Anti-Politics Machine*, proposing that the state, the firm, and the international agency are not the only sites for collaboration. He appeals to academics to think about working with non-state forces and organizations "that challenge the existing dominant order," to look for ways their expertise can be helpful in changing both the field of visibility and perhaps, the structure of eyeballs themselves.

Others make similar arguments. Larry Lohmann (2011) calls for widespread alliance building in support of social movements who are already doing the work of "keeping the oil in the soil, coal in the hole, and tar sand in the land," who are becoming increasingly aligned with those working for biodiversity-rich farming practices and community-based conservation initiatives. Ecologist Claire Kremen (2015) argues that conservationists need to concern themselves with policy reforms that can reduce the advantages to large agribusiness, aligning with major international movements in support of small and mid-sized farmers. I, along with my friends Rosemary Collard and Juanita Sundberg, have called for conservation to embrace a politics of decolonization, one that reckons with "bad decisions," of course, but also aims to reckon with centuries of imperial ruination that have left a world of socioecological asymmetries (Collard et al. 2015). Like Lohmann and Kremen, we think it is crucial to align ourselves with these movements that already exist – like Via Campesina and the self-determination efforts of Indigenous movements – who are already doing the work of not only opposing bad developments and decisions, but also calling elites and institutions into account for ongoing destruction.

Such solidarities are perhaps nonsensical when one thinks of Conservation International or The Nature Conservancy, who as institutions are deeply enmeshed with large multinational companies and often refract the interests of elites. But it is less nonsensical when one thinks about individual scientists and ecologists who are based in academic institutions, or even who are based in these international conservation institutions who are skeptical about enterprising nature as political-scientific strategy.

This, I suggest, could be the living, breathing face of a critical or radical ecology, one built across diverse disciplines (across the humanities, social sciences, and so-called "hard" sciences), conducted in close conversation not to serve elite needs, but to serve movements with a real chance of creating abundant, diverse futures. I think something like these strange alliances – Gretchen Daily and Via Campesina – is what Donna Haraway means in her "Situated Knowledges" essay when she calls for the production of "partial, locatable, critical knowledges" that are "potent for constructing worlds less organized by axes of domination" (1988, 585).

These kinds of knowledge practices could perhaps help manifest the latent radical nature of ecological thinking – an ecological practice and theory that is aligned with the promises of the Enlightenment but also suspicious of reductionism and mastery narratives of Western development and progress (Code 2006; also chapter 2, this book). Situated knowledges, then, are neither only about whom one allies with, nor only about what one sees, but also about how one sees the world. Situated knowledge – or what I am calling a practice of critical ecology – asks that the object of knowledge be seen as something other than a "screen or a ground or a resource" (Haraway 1988, 592) whose rules await discovery. Rather, situated knowledge requires seeing the world as an "actor or agent" (593), as perhaps having "an independent sense of humor," as a "witty agent" or "Trickster," whose codes are "not still, waiting only to be read" (593). This, understandably, may make scientists squeamish or uncomfortable. But for me Haraway is suggesting an onto-epistemology that I found many scientists and ecologists I talked with already adhere to: viewing the earth as a "coding Trickster" is a call to discard dreams of mastery, to embrace highly dynamic, uncertain, and deep unknowns of the future, while at the same time a call to "keep searching for fidelity, knowing all the while we will be hoodwinked" (594). Making room for such a critical ecology within market society, within the contemporary institutions of our time, will not be easy. But market society turns out to be resistant to even the most reductive, economistic science. Could a strange and unkempt alliance of scientists, activists, academics, farmers, Indigenous people, urban residents, and rural people, who not only demand a different vision but also dramatic redistributions in wealth and power, have more luck?

Notes

1 For a wide array of views see Collard et al. (2015), Kareiva et al. (2012), Marvier (2014), Marris (2011), Miller et al. (2014), Robbins (2014), Robbins and Moore (2013), Soulé (2013), Wuethner et al. (2014).

2 Critical investigations include Adams (2013), Adams and Redford (2010), Muradian et al. (2010), Kosoy and Corbera (2010), McAfee and Shapiro (2010), Robertson (2012), and Sullivan (2010a).

3 This chapter draws on 17 interviews conducted in 2008–2010 with experts at the center of ecosystem service science and policy. Where consent was granted I identify interviewees by name; others are identified by their particular job and institutional affiliation. I spoke with people affiliated with the following institutions: World Wildlife Fund (WWF), The Nature Conservancy, Royal Society for the Protection of Birds (RSPB), Packard Foundation, International Union for the Conservation of Nature, the University of British Columbia, Stanford University, Gund Institute Vermont, Duke University, University of California–Berkeley.

4 The editorial was the result of discussions amongst a group of well-seasoned civil society international negotiators.

5 The numbers in the table were obtained by searching the term "ecosystem services" in the decisions of the CBD. The middle column is the number of discrete decisions that mention ecosystem services, no matter how many times it is mentioned.

6 Walter Reid co-wrote the 1990 *World Conservation Strategy* (discussed in chapter 2), and the lead scientist was Hal Mooney, who headed the Scientific Committee on the Problems of the Environment (SCOPE) program on biodiversity and ecosystem function (discussed in chapter 3), and was a colleague of Daily and Ehrlich at Stanford. Their participation in multiple initiatives related to enterprising nature illustrates the small and densely interlinked nature of these circuits.

7 The Ehrlichs' 1981 book *Extinction* put forward the argument that species diversity is central to the earth's life support systems as it provides a basis for healthy ecosystem functioning, which in turn delivers the ecosystem goods and services humanity needs. A history of ecosystem services in the journal *Ecological Economics* (Gómez-Baggethun et al. 2010) also locates the contemporary introduction of the term ecosystem services in *Extinction*, noting that the Ehrlichs use the terminology as a "pedagogic" tool to "demonstrate how the disappearance of biodiversity directly affects ecosystem functions that underpin critical services for human well being" (1213). Gómez-Baggethun and his co-authors (2010) also chart an extensive "pre-history" of the terminology within the Classical economic period to neo-Classical economic period focused on understanding the changing ideas of land and natural capital. In their history of the terminology of ecosystem services, Ehrlich and Mooney (1997) argue that the notion of ecosystem services can be traced back to George Perkins Marsh's (1864) *Man and Nature*, which challenges the idea that natural resources were infinite. Ehrlich and Mooney then point to three key authors – Fairfield Osborn (1948), William Vogt (1948), and Aldo Leopold (1949) – who all promoted recognition of human dependence on the environment.

8   For the early discussions of the term, see Daly (1973) and Odum (1981), and for an economic history see Gómez-Baggethun et al. (2010).

9   Costanza and Daily cite a 1995 meeting they both attended as Pew Scholars as a crystallizing moment for their flagship publications. The Pew Charitable Trusts is an independent, US-based group of charitable trusts, which offers prestigious awards of funds to support policy-relevant research in a number of fields, including on the environment (Pew Charitable Trusts 2013).

10 Conservation investments are also influenced by what Lorimer (2007) calls "nonhuman charisma" – the appeal of animals that are cute, or have features that humans appreciate on an affective, aesthetic basis (e.g. grizzly bears, pandas, tigers).

11 By the turn of the millennium, the lack of economic calculus is a source of critique, as I discussed at the end of chapter 3.

12 For example, the 1990 Clean Air Act in the US establishing a cap and trade system to deal with emissions leading to acid rain. Wetland banking – a scheme that allows for those who impact wetlands to purchase credits from firms who produce functioning wetlands – emerged in the 1980s to meet the broad principles of this Clean Water Act in the US, while not impeding development (see Robertson 2004). Perhaps most famously, the 1997 Kyoto Protocol, which established binding reductions for developed countries, followed in the footsteps of the Clean Air Act by allowing for "developed countries" to exceed their reductions by buying offsets produced in the Global South through the Clean Development Mechanism (CDM). The CDM, while mostly producing credits related to increased efficiencies and fuel switching, does include credits produced via the carbon sequestration services of living systems, defined technically as "Land Use, Land-Use Change and Forestry" offsets.

13 Costanza's $33 trillion estimate came from summing the results of varied individual valuation exercises in a variety of global biomes, and multiplying each estimate by the global acreage of the biome. This is a valuation methodology known as "benefits transfer" because it takes valuations from one site and transfers then over a whole ecosystem. This multiplication of estimated price by quantity is obviously a dramatic simplification of global ecosystemic complexity, and benefits transfer is frequently critiqued on these grounds (see, for example, Nelson et al. (2009)). However, economists' criticisms focused on Costanza's attempt to capture the entire value of the Earth, as opposed to capturing the marginal value of the net lost or gained unit of services provided by the Earth. Economists were merciless in the 1998 issue of *Ecological Economics* devoted to analyzing Costanza's argument (see Costanza et al. 1998). Several interviewees noted that Costanza's approach tainted the concept for several years. One noted in particular that while Costanza's giant $33 trillion amount put the concept on the map, it also "made it so that for ten years it was impossible to talk about ecosystem services with mainstream economists without it being immediately dismissed," creating an "enormous barrier to progress for ten years." Those at the helm of the MA wanted to avoid similar problems by ensuring that mainstream economists were at the center of the project.

14 Many scholars have written about the practices of "fortress conservation," which disregarded the interests and perspectives of local or Indigenous communities and sometimes even resulted in eviction from conservation areas. See, for example, Brockington (2002), Brockington and Igoe (2006), Chatty and Colchester (2002), and Neumann (1998).

15 Political theorist William Chaloupka (2003) describes strategy as the point where the normative comes into contact with the pragmatic; strategy is where political actors are forced to ask questions beyond what is the right or wrong direction; strategy is asking what coalitions one can form, about how others will be convinced to care about something or act differently.

16 In *The Birth of Biopolitics*, Foucault suggests that in liberal society, it is interests that "constitute politics and its stakes" – only when an adequate interest is at stake is a government intervention legitimate (2008, 45).

17 Although there are signs that *proponents* of ecosystem services recognize this conceit; see for example Raymond et al. (2013).

18 For academic papers using InVEST see Arkema et al. (2013), Guerry et al. (2012), Kareiva et al. (2011), Nelson et al. (2009), Tallis and Polasky (2009).

19 Other directors include Jonathan Foley (Director of the Institute on the Environment), Jon Hoekstra (Vice President & Chief Scientist, WWF), Mary Ruckelshaus (Managing Director), Steve Polasky (University of Minnesota), and Taylor Ricketts (Gund Institute for Ecological Economics at the University of Vermont).

20 The Panopticon is a building designed by Jeremy Bentham in the late eighteenth century that allows an observer to observe all inmates of an institution without these inmates being able to tell whether they are being watched. It is a type of architecture that allows for "permanent visibility" to use the words of Foucault (1995) (see note 14, chapter 1).

21 This point was echoed in my interviews with leading ES scientists; as one remarked, "[the MA] didn't do anything in terms of concrete analysis," it didn't do much to "characterize functional forms," it didn't show how "services would be affected by different kinds of management in a quantitative sense."

22 It works by dividing up the land into hexagons, each of which can contain more than one land use or land cover type, but the model uses the the dominant type in mapping. InVEST calculates the implication of the change in each of the hexagons, which they quantify in terms of ecosystem services.

23 The 17 services include blue carbon (carbon storage and sequestration in coastal ecosystems), carbon (carbon storage and sequestration in terrestrial ecosystems), coastal protection; (the benefits of nearshore habitats for coastal protection), coastal vulnerability (the relative risk to coastal areas from storms), crop pollination (the contribution of wild pollinators), habitat quality, habitat risk assessment, managed timber production, marine fish aquaculture, marine water quality, offshore wind energy, recreation, reservoir hydropower production (water yield), scenic quality, sediment retention, water purification, and wave energy.

24 Popper advocates a more modest form of positivism, one based in rigorous hypothesis testing. He was famously critical of attempts to transfer predictive capacities in areas like astronomy to more earthly, complex relations. As he states: "The fact that we predict eclipses does not, therefore, provide a valid reason for expecting that we can predict revolutions" (Popper 2014, 458). The same might be said of attempts to model changes to GDP in 20 years' time based on different land use configurations. Thanks to Vinay Gidwani for alerting me to this distinction.

5

Protecting Profit
Biodiversity Loss as Material Risk

The World Economic Forum (WEF) in Davos, Switzerland is the infamous meeting place of the world's most successful capitalists and world leaders, and perhaps not a site at which one might expect to encounter discussion of biodiversity loss. The annual *Global Risks Report* released at the 2010 WEF, however, identified biodiversity loss as a risk to "keep on the radar." According to the authors – the representatives from business, academia, governments, and international organizations that form the WEF's "Global Agenda Councils" – biodiversity loss had a 15–20% likelihood of occurring, with a potential cost to the global economy of around $30 billion. Figure 5.1 reveals that, in the minds of the WEF authors, biodiversity loss was a risk on par with international terrorism (though slightly more likely to occur). With its fully blocked square on the graph, biodiversity looked to take its place as a global economic concern – albeit a lower-level concern – of the world's most powerful people.

Biodiversity loss, according to the WEF *Global Risks Report*, poses "'material risks" to the global economy. The word "material" has a particular meaning in financial and corporate worlds: if a risk is material, it has the potential to impact the bottom line. International consulting companies now study biodiversity risks; insurance corporations are starting to look for ways to sell new biodiversity-related risk protection products. At the same time, a wide range of international and conservation organizations have also picked up on this idea of biodiversity loss as material risk, including the World Business Council for Sustainable Development (WBCSD), the International Union for the Conservation

Enterprising Nature: Economics, Markets, and Finance in Global Biodiversity Politics, First Edition. Jessica Dempsey.

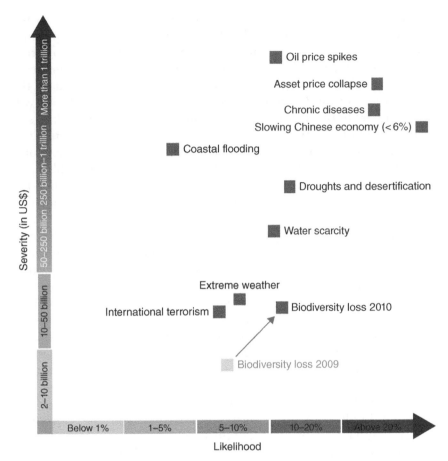

Figure 5.1   Biodiversity in the global risk landscape. © World Economic Forum 2010 (by permission).

of Nature (IUCN), the World Resources Institute (WRI), UNEP's Finance Initiative, and Fauna & Flora International (FFI). Through a variety of research and policy programs, often in collaboration with financial firms, they have sought to represent biodiversity loss as a financial risk that firms will recognize as having an impact on the bottom line, on profits and losses. For conservation organizations, the goal is to produce an ecological risk that "capital can see" (Robertson 2006) – a material risk to which investors will respond – thus also catalyzing investment in conservation.

This chapter is the story of these many actors' attempts to account for biodiversity business risks (BBRs).[1] The transformation of biodiversity loss into a material risk to the bottom line of firms is a crucial part of the project of enterprising nature; it aims to make biodiversity and the currently undervalued "work of nature" permanently visible to the business world. Accounting for biodiversity risks is about the promise of the bioeconomy; it is a tangible site where elites are trying to suture together the ethical and biopolitical interests of nature conservation with financial and corporate logics and practices. Through the identification of BBRs, conservation is to occur naturally. It will become a part of the decisions of individual firms and the workings of markets. Yet "biodiversity loss as risk" turns out to be very difficult to realize; biodiversity loss and ecological change are hard to transform into risks for individual firms or even investors. For the invisible hand to do its work (as the theory goes), the hand needs the right tools – the calculative devices – that can transform ecological, biopolitical risks into financial ones.

In what lies ahead, I explore the proliferation of ideas around this new form of corporate conservation, which is an example of an emerging and hybrid form of knowledge production that I call (following Rajan (2006)) "venture ecology": the increasing entanglement of ecological data and nature conservation with return-generating institutions and initiatives. Making biodiversity loss into a financial risk, as with any enterprising, requires work, new partnerships, new institutional alignments, and, especially, the creation of new calculative devices. In chapter 4 I described another calculative device that "enterprises" – InVEST – a tool produced by ecologists based at both academic and conservation institutions. In this chapter, I introduce two other calculative market devices, the Integrated Biodiversity Assessment Tool (IBAT) and the Corporate Ecosystem Service Review (CESR), both of which aim to render ecological changes economically legible for firms, a specific kind of enterprising. A key premise of this chapter is that these calculative market devices are not simply technical or observational, but are productive as they aim to make biodiversity loss "tractable to the agencies of calculation" (Lohmann 2009, 503) and especially as they attempt to bring biodiversity "into alignment, conceptually, semiotically, and materially, with capital" (Sullivan 2013, 201).[2]

Bringing biodiversity into alignment with capital, however, need not mean new kinds of commodities. Rather, accounting for the risks of biodiversity loss and ecological degradation is less about creating new commodities out of nature, and much more about "reproducing the conditions of production" (Felli 2014, O'Connor 1998). These unwieldy terms have a specific meaning that is important. The "reproduction of the conditions of production," refers to the "perpetuation of the conditions (social, material, environmental, etc.) necessary for the reproduction of *capitalist* social relations" (Felli 2014, 255; my emphasis). BBRs, then, are focused on identifying the socioecological entities necessary to continue fulfilling the needs of a firm, or a group of firms, so they can remain profitable. Imagine, going back to the cover of the book, that it is almond giant Blue Diamond's CEO behind the desk, and payments are being issued to the socioecological creatures who provide the "nature's services" necessary for Blue Diamond to produce and remain profitable. This would involve a whole host of ecological inputs such as healthy soil microbes, but also the work of pollinators, such as bees.

Venture ecology, here, is less about finding new "things" to make a buck off (although that is at play), and more about constructing a smoother space for developments, more about protecting status quo capitalist social relations. And the conservation organizations involved in producing BBRs hope that the idea will simultaneously lead to investments in the work of securing the "reproduction of social life" (Felli 2014, 255), the "necessary conditions for social life on earth, including in, against, and beyond capitalist social relations" (255), meaning that investments by firms to save their bottom line are hopefully also investments in saving the planet, more broadly.

As with the project of enterprising nature in general, the results are murky.

Defining Biodiversity-Related Risk

Ulrich Beck (1992) famously argued that modern society – despite its increasing medical and scientific achievements – is increasingly defined by risks. Paul and Anne Ehrlich's *Extinction*, with its very modernist analogy of individual species as "rivets" on an airplane wing (see chapter 2), was focused on the risk biodiversity loss posed for humanity. Which "rivets" were essential to keeping the whole plane in the air? The Ehrlichs argued that the question was impossible to definitively answer, and therefore every species loss was a risk to the system. The risks with which the WEF and others are concerned, however, are far more specific than the "airplane" as a whole: they are concerned with calculating the impacts of biodiversity loss on firm costs and profitability.

A 2010 WEF report, co-authored by PricewaterhouseCoopers (PwC), *Biodiversity and Business Risk*, defined biodiversity risks as "business risks related to biodiversity in the broadest sense" (WEF 2010b, 2). The report's authors suggest that the risks of biodiversity loss to private firms fall within five categories: reputational, operational, regulatory, market, and financial. Much of the report focuses on the second category – the operational or physical risks of biodiversity loss, including reduced productivity, scarcity of production inputs or increased costs for operations, and disruption of operations.[3] The broader BBRs discourse frequently shares this focus on operational risks and, as a result, seeks to quantify physical risks and offer tools and devices by which firms can account for them. This quantification of physical risk – while extremely difficult – makes some practical sense, as BBRs advocates seek to translate complex ecologies into metrics familiar and legible to the corporate world. I argue throughout this chapter, however, that most of the existing biodiversity loss–related risks to businesses are political, falling largely within the categories of reputational and regulatory risks. These are the risks that have the most frequent and tangible impacts; businesses recognize them most readily as being real and substantive. Overall, the report's authors conclude that no sector "escapes untouched by some form of biodiversity risk" (WEF 2010b, 8). All sectors – including primary industries, utilities, consumer goods, consumer services, health care, industrials, financials, and technology and business services – have some exposure to the risk of biodiversity loss.

At the same time, WEF and others are interested in economic opportunities that might be created by managing biodiversity loss–related risks. Risk is perhaps the most important consideration in private sector decision-making. Capitalist enterprise can even be characterized as an exercise in risk management: "risk taking" is the source of (and justification for) profit. Capitalist firms or individuals invest in an enterprise hoping that the investment will provide a return. The rates of return are most often linked with the level of risk – the higher the risk of the investment, the higher the potential rate of return. Viewed through the eyes of the capitalist, then, risk is linked not only to financial costs but also to financial opportunities waiting to be exploited. As a WBCSD interviewee said, "Somebody's risk is someone's opportunity." Business biodiversity risks are thus a key part of the promise of the bioeconomy, an idealized vision of how biodiversity conservation and capital accumulation can be jointly effected.

Partners in creating BBRs

The idea of biodiversity risk spurs two new linked projects: the search for economic opportunities in the management of biodiversity-related risks and the exploration of new approaches in biodiversity conservation practice. The sets of actors involved in these two projects are distinct – though increasingly and complexly linked – and have, in general, different end goals. The first group, international business organizations and diverse private firms (i.e. World Economic Forum, World Business Council for Sustainable Development, PricewaterhouseCoopers, Swiss Re), work to make biodiversity business risks visible and legible because they seek to protect themselves against lost profits and to source new opportunities for wealth creation. For firms, the BBRs strategy is not one that works to change the operation of capital to accommodate diverse natures; rather, it is a strategy to identify the types of ecologies necessary to ensure the smooth operation of the firm.

The second group, conservation and international organizations like the IUCN, UNEP, FFI, and WRI seek to make biodiversity legible to the private sector; they hope that this legibility will incite private firms to act differently. If firms can clearly see the financial risks of biodiversity loss, the logic goes, they will do whatever is in their power to mitigate them, i.e. reducing biodiversity loss by investing in conservation, thereby facilitating an alignment of biodiversity and capital. For example, if Kraft faces enormous risks from pollinator collapses, the firm will invest in pollinator conservation or work to convince the government to make this investment, or perhaps even advocate for better national and international laws to protect pollinating species.

Conservationists' interest in BBRs is also related to income-earning opportunities for their organizations; as I describe below, they can, for example, sell biodiversity-related information to companies. In this way, the distinction I have drawn between the two groups of BBRs advocates and their aims – while useful at a broad level – grows far blurrier in practice, as part of the "implosions" of venture ecology. In short, though, conservationists' pursuit of meaningful biodiversity business metrics is part of a broader trend, as conservationists attempt to channel and make use of corporate interests in biodiversity. The corporate interest, however, in turn shapes the goals and objects of conservation.

Biodiversity: Risk and the Corporate Interest

Conservationists' new partnerships reveal a well-documented trend in global environmental politics: closer and closer affiliations between environmental organizations and large companies (whose operations often have large-scale environmental impacts). Since the early 1990s, the world's largest multinationals and the world's largest environmental organizations have become increasingly entangled; for example, partnerships exist between the International Union for the Conservation of Nature and Shell, the World Wildlife Fund and Coca-Cola, and Conservation International/The Nature Conservancy and BP (British Petroleum). Observers have studied the rise of corporate conservation (Igoe et al. 2010, Macdonald 2010) and claimed that these alliances are undermining the environmental movement (Hari 2010).

Much corporate interest in conservation is characterized as corporate social responsibility (CSR), the "voluntary contribution of finance, goods or services to community or governmental causes" (Lipschutz and Rowe 2005, 131). However, the inclusion of biodiversity at the Davos meeting represents an attempted departure from CSR – which focused on volunteered contributions for improved public image – and demonstrates a growing belief circulating in the elite corporate and business worlds that there are actual limits to the planet, and, most significantly, that businesses and investments may be threatened by these limits. An editorial in the *Guardian* declares: "Biodiversity and profit can go hand in hand: With the potential for water contamination, drought and soil erosion to impact on future profits, there has never been a stronger business case for change" (Choucroun 2010). The consulting firm McKinsey (2010) states in relation to the issue of biodiversity that the "perceptions of these issues have shifted over the years: executives once approached them purely as a public-relations risk or opportunity but now recognize the real impact they can have on operations and corporate value." The BBRs discourse suggests that – at least to some degree – companies and international organizations are identifying a set of uncertainties regarding how nonhuman nature will affect businesses' bottom line, even outside of market campaigns targeting corporate practices (i.e. beyond reputational risks). The extent of this concern regarding biodiversity-related uncertainty is debatable, as I explain below. Nevertheless, the idea that some CEOs and their companies are cognizant of the material impacts of biodiversity loss is central to the proliferating BBRs discourse: firms are concerned that changing ecologies are threating the reproduction of their "conditions of production."

Indeed, much of this discourse on the risks to companies centers on the internalization of environmental externalities. In a webinar in which I participated in 2010, Ivo Mulder, from UNEP's Financial Initiative (and author of a major report on biodiversity business risks), cited a report that valued global externalities at \$6.2 trillion a year and argued that these externalities – a cost or benefit not included in the price of goods and services – present a "huge time bomb for investors because these costs will, in the end, get back in the balance sheet." Similarly, authors of the WBCSD report *Vision 2050* envision a radical new landscape for business, one "based on a global and local market place with "true values and costs," the "truth" being established by the limits of the planet and what it takes to live well within them" (WBCSD 2010b).

But making biodiversity loss legible in financial risk terms is not automatic or evident in existing practices. Producing biodiversity loss as material risk, as with enterprising nature in general, takes coordinated action that draws together the interests of heterogeneous actors and groups: consultancies, governments, international organizations, universities (especially business schools), investment firms, and NGOs. The result is a proliferation of research and report writing; tangible changes to business practices remain elusive or at least difficult to document materially.

Research and reporting

In 2004, ISIS Asset Management – a UK-based company with £62 billion under management – collaborated with the Earthwatch Institute to produce a report: *Is Biodiversity Loss a Material Risk for Companies?* The report argues that yes, "biodiversity does present material risks to many companies in which ISIS invests, although putting a precise financial value on such risks in not often possible" (ISIS 2004, 4).[4] The report was the earliest sign of corporate attention to something called "biodiversity risk." It concluded that companies were not managing biodiversity risks effectively, and "therefore, this report should serve as an early warning to companies ... [to] start investigating and managing biodiversity-related risks now" (ISIS 2004, 4). (The report, though, was funded by the UK Department of International Development, a fact that suggests that even as ISIS argued that biodiversity loss was a material risk, the threat did not warrant much investment of its own substantial resources!)

Following the production of this ISIS-Earthwatch report, other firms, environmental organizations, international organizations, government agencies, and schools further elaborated on the notion of BBRs (Table 5.1). Cross-sectoral and multi-site collaborations often granted their reports legitimacy and authority. Most of the institutions focused on BBRs are located in Europe, particularly in the UK, the Netherlands, and Switzerland. A large contributor to the BBRs discourse is UNEP Finance Initiative (UNEP FI) (which is located not in UNEP headquarters in Nairobi, but in Geneva); the group holds the legitimacy of a UN organization, even though most of its work is funded by the private sector, which, according to one interviewee, "has a big say in how UNEP FI approaches the issues." Igoe et al. (2010, 490), drawing from Sklair (2001), refer to this conglomeration of groups as a Gramscian "sustainable development historic bloc," a transnational capitalist class of "corporate executives, bureaucrats, and politicians, professionals, merchants and the mass media." However, while these reports might be the outputs of elite transnational bureaucrats and professionals, I want to emphasize that their message is still marginal to mainstream private sector decision-making (as I soon explain); firms are, by and large, not considering biodiversity loss within their risk assessment frameworks.

Table 5.1 Reports addressing the risks of biodiversity loss

Year	Report title	Report author	Collaborating institutions	Donor(s) or Sponsor(s)
2004	Is biodiversity a material risk for companies	ISIS Asset Management (now F&C) – UK based	Earthwatch Institute (NGO)	UK Department of International Development
2007	Biodiversity, the next challenge for financial institutions? A scoping study to assess exposure of financial institutions to biodiversity business risks and identifying options for business opportunities	Ivo Mulder, International Union for the Conservation of Nature (IUCN)	Alcoa Foundation (Foundation of Alcoa, aluminum producer), Alterra (Netherlands-based research institute)	Alcoa Foundation
2008	Dependency and impact on ecosystem services – unmanaged risk, unrealised opportunity: A briefing document for the food, beverage and tobacco sectors	The Natural Value Initiative (collaboration between NGO Fauna & Flora International, Brazilian Business School FGV, and UNEP FI)	The Natural Value Initiative is guided by a steering committee that includes many international financial institutions, NGOs, and consulting firms.	Netherlands Ministry of Housing, Spatial Planning and the Environment (VROM)
2009	The global state of sustainable insurance: understanding and integrating environmental, social and governance factors in insurance	Insurance Working Group of UNEP FI (includes Swiss Re and Allianz)	Fox School of Business at Temple University	UNEP FI receives contributions from participating financial institutions

Year	Title	Organization	Contributors	Funding
2010	Biodiversity and business risk	World Economic Forum commissioned, produced by PricewaterhouseCoopers (PwC).	N/A	The World Economic Forum is funded by its 1000 member companies
2010	The next environmental issue for business	McKinsey and Company	N/A	McKinsey and Company
2010	Demystifying materiality	UNEP FI	Representatives from nine different organizations are listed as contributing authors, including Credit Suisse, Vic Super, and Rabobank (for example)	UNEP FI receives contributions from participating financial institutions
2010	'COP' Out? Biodiversity loss and the risk to investors	EIRIS	N/A	EIRIS Foundation
2011	Sustainable insight: The Nature of ecosystem service risks for business	KPMG	UNEP FI, Fauna & Flora International	Internally funded by organizations
2011	Biodiversity principles: Recommendations for the financial sector	Association for Environmental Management and Sustainability in Financial Institutions (Germany)[5]	Centre for Sustainability Management (CSM) at the Leuphana Universität Lüneburg (Germany), UNEP FI, PricewaterhouseCoopers, Deutsche Bank, and others.	German Federal Agency for Nature Conservation

(continued)

Table 5.1 Reports addressing the risks of biodiversity loss (*Continued*)

Year	Report title	Report author	Collaborating institutions	Donor(s) or Sponsor(s)
2012	A new angle on sovereign credit risk: Environmental risk integration in sovereign credit analysis	UNEP FI	Global Footprint Network	UNEP FI receives contributions from participating financial institutions
2014	EU Business and Biodiversity (B@B) Workstream 1: Natural capital accounting for business.	James Spurgeon, Sustain Value, commissioned by EU Business and Biodiversity	EU Business and Biodiversity, ICF International (consultants), with support from nine companies including British American Tobacco, Lafarge, and Heidelberg Cement (for example)	EU Business and Biodiversity coalition
2015	Making the invisible visible: Analytical tools for assessing business impacts & dependencies upon ecosystem services	Business Social Responsibility Network (BSR)	This report emerged from the BSR Ecosystem Services Working Group. There are 12 corporate participants including Chevron Corporation, the Dow Chemical Company, Exxon Mobil Corporation, and the Walt Disney Company (for example)	Business and Social Responsibility Network (network funded by over 250 corporate members)

Within a few years of the 2004 ISIS-Earthwatch paper, reports on BBRs – almost always arising from cross-sectoral partnerships – began to appear more steadily. A 2007 IUCN report, written by Ivo Mulder (now the coordinator of the Business and Ecosystem Services "workstream" at UNEP FI) with support from Alcoa Foundation (foundation of the international aluminum producer Alcoa) and Alterra (a Netherlands-based research institute), argued that BBRs could be the next challenge for financial institutions (Mulder 2007, xi). A 2009 UNEP FI report, *The Global State of Sustainable Insurance*, focused on the need for insurers to incorporate biodiversity and ecosystem risk factors into their underwriting.[6] (The report was produced by the Insurance Working Group of UNEP FI, which includes the world's major insurers and re-insurers, such as Allianz, Swiss Re, and Lloyd's, in collaboration with the Fox School of Business at Temple University.) This work culminated in the 2010 WEF–PwC report (WEF 2010b) *Biodiversity and Business Risk* that brought the notion of biodiversity to the world's elites. Another 2010 report by global consulting giant McKinsey surveyed over 1500 executives about "what biodiversity means, how important it is to their businesses, and why" (McKinsey 2010). The title of a *Financial Times* story on the McKinsey research gave the gist of this research and reporting trend: "Saving Species: Bad for Biodiversity Is Often Bad for Business" (Harvey 2010).

That these reports share many common characteristics is evidence of the circulation of ideas through linkages in the "sustainable development historic bloc" (Igoe et al. 2010, drawing from Sklair 2001). The reports deploy a similar categorization of risk types (i.e. operational, regulatory, market, reputational, and financial) and cite biodiversity loss–related statistics, often drawn from the *Millennium Ecosystem Assessment* (MA 2005). The reports include, in boxes, case studies of lost shareholder value, and even the specific examples are repeated:

- Britain's largest port operator, Associated British Ports, lost £155 million in market value when the government blocked its plans for a new container terminal due to potential impacts on marine wildlife;
- the 2007 collapse of bee colonies cost US producers $15 billion due to low pollination;
- the pipeline company Transneft faced costs of $1 billion when it had to reroute a pipeline due to public outcry over the sensitive ecosystems it would potentially destroy in Russia.

These reports also share – again in textbook-like boxes – declarations from corporate actors about the significance of BBRs: a Kraft executive notes that colony collapse disorder is contributing to a reduced availability of almonds (WEF 2010b); the CEO of Hermes Equity Ownership Services declares that the loss of natural capital (including biodiversity) has "direct and widespread negative effects on financial performance" (UNEP 2010, 5). In making the case for corporate attention to BBRs, the reports emphasize increased costs or interrupted production. Taken together, they reveal a densely interlinked and highly repetitive corporate–conservation effort to call businesses' attention to the material risks of biodiversity loss and to the eroding conditions necessary for capitalist production.

The Challenge: Biodiversity Loss Is Not a Material Risk, "Naturally"

Yet the arguments and exhortations of these reports are sharply disconnected from the present-day priorities and realities of businesspeople and the financial world at large, wherein relatively little concern exists regarding ecosystems and biodiversity. Biodiversity loss may possibly be a concern for some CEOs; a survey in the WEF report found that 27% were extremely or somewhat concerned (WEF 2010b). A 2009 survey, however, found that only two of the 100 Financial Times Stock Exchange (FTSE) companies recognized biodiversity to be of strategic importance to their businesses (TEEB 2009). The McKinsey (2010) survey found that only 12% of executives and managers projected a significant operational risk resulting from biodiversity issues, with many saying they face no risks at all. A separate study found that only 6 of 50 banks have taken steps to account for biodiversity-related risks and opportunities (Mulder and Koellner 2010). An Experts in Responsible Investment Solutions (EIRIS) analysis of 1800 publicly listed companies found not only that very few had adopted policies on biodiversity, but also that sectors with high biodiversity impacts in their supply chains are failing to address these impacts.[7] Given this low level of concern, it is unsurprising that few companies have taken steps to address biodiversity loss as either a risk to their operations or as a site of new financial opportunities.

But if biodiversity loss poses a material risk to a wide range of sectors – as argued by so many actors in the reports above – why do most businesses and managers show so little concern?

A key reason is that business biodiversity risks are, it turns out, are very difficult to ascertain in terms meaningful for corporate decision-makers, despite the efforts of these organizations and report writers. The impacts of biodiversity loss are difficult to quantify for several reasons. First of all, the effects of these impacts are cumulative and therefore hard to attach to specific actions or effects. "The effects of biodiversity loss are not, in most cases, dramatic one-off events," reports the WEF (2010b, 6), "but rather they accumulate gradually, sapping the productive capacity of the economy, and so are less visible to business leaders and political decision-makers." Biodiversity loss is a systemic risk, defined as "the risk of collapse of an entire financial system or entire market, as opposed to risk associated with any one individual entity, group or component of a system" (WEF 2010b, 10). Systemic risks, at some level, have material consequences for firms, but these are highly unpredictable and sudden risks, and as the WEF recognizes, these are not risks that can necessarily be easily translated into specific risks for firms or investors.

In a presentation at the London Biodiversity and Ecosystem Finance conference in 2008, one representative from a large UK-based asset management firm quoted the MA to stress this aspect of biodiversity risk: "Businesses cannot assume that there will be ample warning of a change in the availability of key services or that the company's past responses to change will be successful in the future. Ecosystems change in abrupt, unpredictable ways." Compounding all these problems of ecological unpredictability and indeterminacy is a lack of knowledge regarding species' functions within ecosystems. The impacts of the loss of specific species are difficult to determine; as I described in chapters 2, 3, and 4, ecologists know little about which species (or assemblage of species) are most critical for the ecosystem function and services. As such, it is very difficult to describe to a chief financial officer the risks the firm will face from a specific change in land use or biodiversity composition.

Actors involved in BBRs research and reporting face a significant challenge: how to turn a globalized, systemic, diffuse (and therefore difficult to quantify) risk of biodiversity loss into an individuated risk that can be measured, managed, and mitigated by single firms. As the Head of UNEP FI Paul Clements-Hunt explains: "The rapidly declining levels of biodiversity and associated loss of ecosystem services is clearly significant for society as a whole. However, as yet, this significance has not been translated to material risk exposure on a company level" (quoted in Griggs et al. 2009, 9). Measures that will be meaningful for financial decision-making are extremely challenging to produce: it is "difficult at present to link BBRs to tangible financial metrics, such as market capitalization, asset value or credit risk," notes the IUCN report (Mulder 2007, xi).

This problem leads BBRs advocates to pursue the development of new tools to help financial institutions and corporations discover their true risk exposure from ecological decline. As one interviewee from the private sector arm of the World Bank, The International Finance Corporation (IFC), explained: "You can make a material risk for water [and also] for certain biodiversity-related matters in the agricultural landscape ... We are working on studies to try to establish these things more precisely ... We don't have yet completely this information. But we are working on it." In short, this interviewee said, BBRs is "not yet fully ready and mature ... but it's going to be ... We are getting there on some aspects." More tools are needed to show firms precisely how they stand to lose (or win) due to changing biological diversity, tools that are even more specific than InVEST. This interviewee went on to mention two tools in particular that she was excited about: the Integrated Biodiversity Assessment Tool (IBAT) and the Corporate Ecosystem Service Review (CESR) (both described below).

In spite of a lack of corporate concern and meaningful metrics of accounting, the "biodiversity loss as material risk" concept is rising in conservation discourse, in part because of its appeal to NGOs trying to convince capital to "do the right thing" and account for biodiversity. I said previously that the BBRs discourse could appear to be out of step with realities of the corporate world; however, the discourse is not willfully inattentive to current facts or naively optimistic. It is not opposed to reality. Rather, we might say, as Kaushik Sunder Rajan explains in relation to his study of genomics and drug development, it is promissory, not yet linked to "commercial realization" (2006, 113). BBRs are a "discursive mode of calling on the future to account for the present" (116), establishing new grounds on which the future can unfold. The task is to create the conditions – the "new grounds" – to make biodiversity an entity of permanent visibility in the calculations of firms.

Before I turn to the calculative devices doing this work, I want to draw out a tension between attempts to make "operational risks" from biodiversity loss appear and the actually existing way that civil society organizations, social movements, and governments are already doing the work of making biodiversity loss material.

Strong Movements and Legislation = Biodiversity Loss as Reputational and Regulatory Risk

The reports in Table 5.1 (above) draw out that the most identifiable risks to firms from biodiversity loss are social and political, not operational; even the reports promoting the BBRs approach recognize that the most tangible current biodiversity-related risks to firms' profits come from political organizing and government regulation. In fact, governance-related risks form a line of questioning that runs throughout the above-described reports. "Biodiversity loss as a material risk" is closely linked to the efficacy of advocacy efforts, government regulation, and legal frameworks. For example, McKinsey (2010) survey respondents said that consumer action (37%) and regulatory action (36%) would provide the main impetus for them to take action on the issue of biodiversity.

Consumer action is a significant reputational risk for firms. One interviewee from UNEP FI noted that a key underlying motivation for banks to be involved in identifying biodiversity risks is directly related to consumer campaigns that make the broader public aware of the negative actions of firms:

> Indeed campaigns against projects and initiatives cause a very, very big impact. ... I do believe that both for, say, companies operating in tar sands like the oil and gas sector as well as for those who are actually financing these operations that they increasingly know that there's a Big Brother watching them. And now that we are living in a globalized world it's very easy for them [NGOs, other critical social movements] to get their data and knowledge very quickly and to disseminate that rather quickly and effectively to a large public and scrutinize these companies very, very directly. And so ... it might be one of the key factors why companies are actually doing something about biodiversity.

This statement suggests that, at least for some firms, establishing biodiversity loss as a risk requires intensive NGO campaigns that target the firms' products or developments. For example, the forestry company MacMillan Bloedel suffered reputational damage when Greenpeace and other NGOs protested the firm's clear-cutting practices (WEF 2010b). The protests caused Scott Paper and Kimberly-Clark to stop sourcing fiber from the company, resulting in a loss of 5% of its revenue almost overnight (WEF 2010b). Public and often confrontational action in defense of biodiversity remains a very real and very visible material risk for companies.

Regulatory risks are complex and often contradictory, as government regulations or re-regulations can both enhance and reduce risk. Governmental interventions can have negative impacts on company profits, as was the case for Associated British Ports (described above) when the government blocked development of a new terminal. In another example, the corporation Airbus experienced negative impacts from new European Union regulations. As the company vice-president explains: "When we were developing the A380 assembly line, our obligations under the 'Natura 2000' environmental regulations in France resulted in a six-month delay for one of the buildings while preservation of local biodiversity was addressed " (WEF 2010b, 9). Because of the economic risks of government action, business-oriented organizations such as the

WBCSD state that regulation may be "necessary in certain instances" (WBCSD 2010b, 20), but should be used secondarily – only after policy approaches based on the "realignment of economic incentives" are considered. Furthermore, the WBCSD notes that regulation should not "stifle" the "innovation often provided by voluntary standards and industry initiatives" (WBCSD 2010b, 20).

However, some BBRs advocates do, in general, believe that laws and policies that hold firms accountable for damage to biodiversity (i.e. liability laws) are needed. The WEF survey (2010b, 7) found that 44% of CEOs surveyed believed that national governments are not doing their jobs related to biodiversity and ecosystem protection, "implying a need for more direct government action to address biodiversity loss." Insurance companies participating in the UNEP FI (2009, 14) *Sustainable Insurance* survey stated that environmental risks are "outrunning the development of prudential regulatory or legal frameworks," noting that "regulation is the number one factor influencing underwriting, and the number one factor in terms of risk severity." Government regulation is a crucial factor in making biodiversity loss a more legible – and thus accounted for – risk for firms.

Since many working in the area of finance and biodiversity are interested in regions of "high biodiversity" in the Global South, this raises a vexing problem. If laws and regulations are critical to reducing risk, then what happens in countries or regions where government institutions and enforcement capacities are low? For many advocates of the financialization of conservation, the answer lies in having financial institutions take over government-like roles, especially in the Global South. As one interviewee at UNEP FI said, "Given that fact that most biodiversity is located in emerging economies and developing countries with very weak government institutions and weak enforcement, having financial institutions actually taking over the role of the government by putting restrictions and certain conditions on loans or equity investments – you could actually make a big difference." The Equator Principles aim to do just this by asking signatory financial institutions to assess impacts on "critical habitats with high biodiversity value." In what they call "Non-Designated Countries," those without "robust environmental and social governance, legislation systems and institutional capacity designed to protect their people and the natural environment" (Equator Principles 2013, 15), the principles require banks to evaluate project impacts in relation to not only domestic laws and policies but also the International Finance Corporation's Performance Standards.

In many ways the interviewee demonstrates what James Ferguson describes as the "fraught" relationships between private entities (like conservation organizations, development institutions, and lenders) and some Southern, particularly post-colonial states. These relationships are fraught, Ferguson argues, because the growth and expansion of NGOs, development institutions, and firms in the Global South, which bring capital and technical expertise, are also bringing into being a kind of privatization of sovereignty (cited in Igoe et al. 2010, 496).

Advocates of BBRs hold mixed and contradictory positions on the relation between regulation and risks. On the one hand, proponents of biodiversity business risks suggest that increasing regulation and creating strong governments are crucial parts of creating robust risks that can change the actions of firms and "save nature." On the other hand, BBRs advocates believe that finance itself has a critical role to play in governance, particularly in places without strong regulatory environments and where weak regulatory environments are put forward as justification for firms to take on stronger governance roles.

This is the contradictory but also continuing face of what we might call eco-imperialism today: in the North actors use the formal state processes to produce domestic environmental laws and policies, and then use their superior purchasing power or financial clout to influence the production of nature elsewhere (in the Global South). Some get (still very imperfect) liberal democratic governance, others get a privatization of sovereignty via the governing apparatus of corporate and financial capital.

Calculative "Risk" Devices: New Tools for Risk Production

Even if the most robust risks come from regulatory or political processes, much attention in the world of BBRs focuses on making firms aware of the broader array of risks emerging from biodiversity loss. And more than anything else, making these risks "real" requires new knowledge and methodologies. As an International Finance Corporation interviewee explained, new tools are necessary to "formalize the way that biodiversity is measured and priced on the risk side." She went on to explain that this is a difficult proposition: "It's not that easy, and we're not at the point where we know exactly that this biodiversity, or this risk aspect, equals X amount of exposure."

International groups like UNEP FI and the Business Social Responsibility Network (BSR Network) (a network of 250 companies seeking "sustainable business solutions")[8] have been strong voices in this call for rapid ecosystem and biodiversity assessment tools that can help the private sector "identify, measure, and track" the changes in natural systems, and understand the relevance of these changes to corporate decision-making and risk assessment processes (BSR 2010; UNEP FI 2007). The BSR

Network reports that "[s]imilar to how financial analysts tracking flows of capital during the Great Depression realized that their models were missing underlying indicators of market performance, today's environmental thought leaders recognize that they are missing critical information that could help them understand what enables natural systems to function in expected ways" (BSR 2010). A similar view was expressed at the first meeting of UNEP FI's Biodiversity and Ecosystem Services Workstream; the minutes note the dearth of tools that could tie risks of biodiversity loss to companies (UNEP FI 2007).

What is needed to create robust biodiversity business risks is not general knowledge of systemic risk found through global species counts and ecosystem service declines, but rather more finely tuned, individuated knowledge that links potential profit and loss with ecological predictions and assessments. InVEST, recall from the previous chapter, is a calculative device that aims to produce quantitative assessments of ecosystem services under different scenarios to inform governance; these devices are meant to make risks "imaginable and governable" (Johnson 2013, 36). More generally, Sian Sullivan refers to the importance of what she calls the "ecoinformatics" needed to "measure, assess, standardise and disaggregate" (2013, 205) nature's goods and services (2013, 205). Calculative risk devices are crucial for making biodiversity economically legible in terms of financial risk.

We are well into a methodological boom in calculative devices for assessing and valuing biodiversity loss and ecosystem services. In 2008, the BSR Network published a report comparing seven different tools that assess multiple ecosystem services (BSR 2008), and in 2015 it produced a catalog listing over 120 tools for *Making the Invisible Visible*, the title of its report. In the following two sections, I describe two risk assessment calculative market devices – the IBAT and the CESR – which are intended to make estimations of biodiversity business risks for private sector actors – companies, investors, and lenders. They aim to produce legible biodiversity and ecosystems that are linked to a potential monetary loss or gain. Why these tools? Both were mentioned in my research interviews as tools that are helping make risk a reality, and they are also being used in due diligence processes at banks, within the assessments of project financiers and corporations.

The Integrated Biodiversity Assessment Tool: Avoiding "wild things," avoiding fracas

The Integrated Biodiversity Assessment Tool (IBAT) was created out of a partnership involving BirdLife International, Conservation International (CI), IUCN, and the UNEP World Conservation Monitoring Centre. It is a tool meant to inform the private sector about biodiversity business risks in project planning and for investment risk assessment. On a fee-for-service basis, IBAT gives business access to the world's most comprehensive information about high biodiversity sites in order to inform corporate decision-making. The service helps businesses see "Where on earth is biodiversity" (UNEP 2010), or, as described in the *Economist*, it shows business "where the wild things are" (*Economist* 2008).

The tool brings together a large amount of data from conservation organizations. It enables a company at the project planning stage to log into the website and explore the high biodiversity locations in and around the project's proposed site. As an article in the *Economist* (2008) describes, the purpose of IBAT is to make it easier for businesses to incorporate concerns about biodiversity into their planning from the beginning of a project, not simply when protesters show up at their offices. The article suggests that such a tool might have saved the Russian pipeline firm Transneft the $1 billion it cost to shift their pipeline plans after persistent protests about the sensitive ecosystems around the original route (repeating an example found in other reports). Critically, the website is anonymous; no records are kept of which areas companies look up. The article notes: "BP, for example, says it could use IBAT to check whether areas where it might bid for exploration licences are ecologically sensitive, without alerting competitors to its interest. It also thinks IBAT could come in handy when planning the routes of pipelines" (*Economist* 2008). This, then, is a tool to help firms reduce their exposure to civil society uncertainties, to protect them from political backlash that impacts permitting and can even put the kibosh on projects entirely.

Further, as an IFC interviewee described to me, IBAT is being used by firms, and is helping companies to "tick boxes":

> So when a bank wants to invest in a company that is in the commodity sector, for example, more and more banks need to know where they sit, and they need to be able to tick a box on whether it's a high risk, not high risk, etc. And on the biodiversity side, it was impossible, because nobody could measure anything, and no one could tick boxes.

In other words, IBAT is a tool that helps financial firms understand the risks to a company that an investment in, say, a major infrastructure project might pose due to impacts on biodiversity or ecosystems; the tool lets investors see firsthand if a loan they are making has risks for being close to or within a "high value conservation area" and thus may see difficulties in approval or from civil society backlash.

But what spaces are identified as "high priority" or high value conservation sites by IBAT? First, there are legally defined protected areas, the data for which is drawn from the World Database of Protected Areas (WDPA), which is jointly managed by UNEP World Conservation Monitoring Centre (UNEP-WCMC) and IUCN. Second, IBAT draws attention to what it calls globally important sites for biodiversity. Data for these sites is drawn from BirdLife International's Important Bird Areas, PlantLife International's Important Plant Areas, IUCN's Important Sites for Freshwater Biodiversity, and sites identified by the Alliance for Zero Extinction. The datasets are taken from the World Biodiversity Database (managed by BirdLife International and Conservation International) and are also informed by IUCN's Red List of Endangered Species. Overall, IBAT integrates datasets from many international and conservation organizations (and their corporate partners) into a program that identifies "high risk" sites of biological richness.

The many organizations involved believe that IBAT will cause companies to reconsider their project locations and investments, and that the tool will also be used by investors or banks in their risk assessment protocols. For example, one can imagine the boxes to be ticked by the manager of a due diligence process: "Have you consulted the IBAT tool? Is your project within 20 kilometers of a 'high priority biodiversity site'?" As one IFC interviewee noted, IBAT "overlaps all of the systems to figure out what you have to avoid," meaning it identifies the patches of earth that are considered "high value" for conservation. She went on to say:

> In the end, it's all about whether the companies are protecting biodiversity, and are doing the right thing, having the management systems in place, and having corridors. When you're talking about commodity production, we want to see protected areas, buffer zones, corridors etc., in the landscape.

And especially, for this interviewee, the proper use of IBAT will mean that there is "no more deforestation of critical habitats, and so on and so forth," because firms will relocate their activities to non-critical spaces.

What does this tool produce? Proponents believe that the tool produces, ideally, spaces of "biodiversity risk" for investors, companies, and lenders. It produces a God's eye view of the most critical biodiversity spaces, generating "No Go" zones for business. At the same time, the tool effectively produces "Go" zones, a whole world of land and waters, species and spaces, where businesses may not have to worry about biodiversity issues. This is another mapping of usable and unusable space, albeit the demarcation of one or the other will depend very much on the business sector. IBAT produces a different strand of terra nullius, or empty lands, suggesting that these lands are empty not of people, but rather of biodiversity of significance (as the tool defines significance).

In this way, the IBAT tool not only means to spare some areas from industrial projects, but also helps to create a smoother space for capitalist development in others; it is a tool to help firms avoid the fracas and delays that happen when a firm proposes a pipeline through a high value conservation area.

Within a so-called green economy, however, the "No Go" high biodiversity sites might also be lands of economic opportunity for another firm. Data on these sites can provide direction for potential investments in biodiversity offsets or can enable other bioeconomy-based businesses like forest carbon offsets. These high biodiversity sites might also, of course, become locations for bioprospecting or the development of sustainably harvested products. In other words, IBAT, in bringing together previously disparate information on global biodiversity hotspots in a "single, accessible source" (*Economist* 2008), produces certain spaces as open for business as usual, for extractive industries especially, and other spaces as open for green opportunities. In its attempt to individuate and spatialize biodiversity business risks – i.e. show firms exactly where on earth they might face risks – the IBAT tool helps create spaces of investment and disinvestment.

Venture ecology

The IBAT tool also creates a new business for conservation organizations, as the dataset is accessible only via payment. Companies with annual revenues greater than US$ 1 billion can subscribe annually for US$ 35000, plus applicable taxes. "Ensure your company has the most up to date data sets," the webpage invites. For companies with smaller revenues, there is a "pay as you go" data download subscription, which allows a company to download the spatial data at a cost of US$ 2950 per 10000 km², plus taxes. Firms can also commission a report from IBAT that details key protected areas and biodiversity areas within 50 km² of a location, as well as a list of globally threatened species that may be found near the location, for the low-low cost of US$ 750 (plus taxes).

What are we to make of this strange scenario? Non-profit conservation organizations are finding new ways to support their expensive data collection and management efforts, finding new sources of revenue by enclosing their data from free use by businesses. Further, the IBAT tool is integrated into the Proteus program, which is a partnership between businesses (including Rio Tinto, Shell, BP, Exxon Mobil, and Microsoft) and the UNEP World Conservation Monitoring Centre (UNEP-WCMC). The purpose of the Proteus program is to make available global information on biodiversity, especially to inform extractive industries of biodiversity risks. The IBAT's integration with Proteus is exemplary of venture ecology, a kind of scientific–market collaboration that is trying to achieve two heterogeneous aims at once: the NGOs want to create "zones of no go" to save diverse natures (and make a little money for their labor-intensive data collection efforts) and the firms want a smoother development trajectory for their pipelines and mines. IBAT, then, is a tool that tries to stitch together the "nature-loving" and global social reproduction aims of the participating NGOs with the predominant exchange value concerns of firms, a reoccurring attempt that characterizes global biodiversity conservation from its beginnings (see chapter 2).

Venture ecology means two interlinked things. It refers to new revenue streams for conservation organizations as they provide information to firms about politically risky spaces on the earth (via subscriptions to IBAT). But it also refers to a form of information and knowledge about ecological conditions that will help firms secure their status-quo profitability – ecology to help secure the necessary conditions for capitalist production, including the social license to operate.

It is unclear what these economic arrangements mean for the kinds of data collection and science that are undertaken in the first place by these conservation organizations. Is this implosion of the corporate and scientific/conservation worlds reconfiguring "fact production" itself? Is this emerging "venture ecology" changing the very nature of how knowledge is circulated and produced by NGOs? To be clear, access to the data in IBAT is only enclosed for firms, as researchers and planners are able to use the dataset free of charge in a separate website that requires users to agree to non-commercial use terms. There are signs, however, that, through partnerships like the IBAT, data collection priorities may be increasingly shaped by corporate interests. In an instance of controversial journalism, reporters from the British magazine *Don't Panic* posed as executives from the arms manufacturer Lockheed Martin seeking help with the company's corporate image. They set up – and secretly recorded – a meeting with Conservation International.[9] The CI representative explained how Lockheed Martin could "buy into" the IBAT for $25 000; as a result, the company could be associated with the tool. No regulation of the company's activity would be implied. The CI representative then notes that Lockheed Martin could also join the Corporate Consultative Group for the IBAT, which would allow the company "to potentially craft what other data … may be incorporated in the future."

The fake interview went viral, shared especially among conservationists but also going beyond to news sites like the *Huffington Post*. For some it demonstrated the very low terms on which corporations with spotty backgrounds are invited into partnerships with NGOs. In terms of venture ecology, the fake interview demonstrates how businesses and especially multinational corporations are being invited to help shape what environmental data is collected and made available in tools like the IBAT. Indeed, businesses are invited to become what they now call "Data Development Partners" with IBAT, where firms can, for an unspecified fee, directly "improve data holdings for a specific part of the world," as the website reads. This corporate involvement may be changing the ways in which environmental knowledge is produced and legitimated within conservation organizations, at least influencing the patches of the planet that become sites of more or less intensive ecological monitoring according to corporate interest.

Accounting for risks, enterprising nature: Corporate ecosystem service review

While the IBAT tool focuses on alerting private sector actors to particular "biodiverse" spaces on the globe that they should avoid (to avert risk to reputations, or avoid fracas and unnecessary delays), the Corporate Ecosystem Service Review (CESR) is a tool that aims to help companies identify operational risks and dependencies on ecosystem services, to show them nature's enterprising nature in relation to the needs of the firm. This is a tool that is meant to make firms aware of not the punitive risks of civil society movements, but the risks of declining ecological health.

Developed in 2008 by WBCSD and WRI (notably the same collaboration that produced the widely used Greenhouse Gas Accounting Protocol standards), the tool emerged directly out of the findings of the Millennium Ecosystem Assessment (MA). The MA demonstrated for one WBCSD interviewee "how bad things are" and inspired the development of a risk management framework that could help firms begin to account for often-unrecognized ecosystem risks, dependencies, and opportunities. The CESR, one of the tool developers told me, "was an attempt to provide firms with a flexible, easy-to-use tool that could [be] applied to any sector, any part of a company, to help them figure out what ecosystems services matter to them, and how to respond to that and develop strategies." The tool is rapidly moving around the globe and WRI has a growing database of experts and consultants set up to facilitate these assessments throughout the world, as well as to conduct workshops on the tool.

The methodology for assessing risks involves a series of structured questions through which companies assess their dependence and impact on 28 ecosystem services, in an easy-to-use Excel spreadsheet. These include provisioning services like food, fiber, biomass, and genetic resources; regulating services like climate/water/disease regulation, water purification, and pollination; and cultural services such as recreation and tourism. To assess dependence on each ecosystem service, companies are asked: "Does this ecosystem service serve as an input or does it enable/enhance conditions for successful company performance?" And if it is an important input for performance: "Does this ecosystem service have cost-effective substitutes?" The methodology asks companies to evaluate their impact on these services by posing three questions: (1) whether the company affects the "quantity or quality of the ecosystem service"; (2) whether the company's impact is "positive or negative"; and (3) whether the "company's impacts limit or enhance the ability for others to benefit from the ecosystem service." This final question related to *benefits* is key to identifying potential opportunities, as the CESR directs companies to "[l]ook for ways to monetize ecosystem services the company already provides without compensation" (CESR 2008, 28). Based on this assessment, the tool directs companies to analyze trends in the ecosystem services that they depend on, impact, or provide. It also suggests that companies seek outside information and advice. Finally, the tool directs companies to identify their business risks and opportunities based on the above information.

When explaining how this tool differs from other corporate risk assessment procedures, a tool developer noted that many due diligence systems and environmental management systems "don't focus enough on the dependencies that the companies have on the natural environment, and on ecosystems services." The interviewee continued:

> Some companies look at water, for example, or look a little bit at their supply chain, but not in a systematic or rigorous way. And we've been actually surprised by a number of manufacturing firms who have been working in the same watersheds for years, you know, twenty, thirty years, and as part of their review that they do every two years, a couple of them decided to do ESRs alongside of that, and they said, "Wow, the things we're just not looking at." And it wasn't a problem forty years ago, but we've had so much population growth, and we've got climate change now, and things around us are changing and we're just not paying attention to it.

Where's biodiversity?

It is important to note, however, that the CESR approaches the problem of biodiversity loss through the lens of ecosystem services. Indeed, the tool does not even list biodiversity as one of the 20 ecosystem services, and thus does not direct companies to assess their dependence, impacts, and opportunities from biological diversity. When I interviewed tool developers at WRI and WBCSD and corporate road testers, each insisted that biodiversity is implicit in ecosystem services, but that biodiversity is not easily understandable in the corporate world. "We don't like biodiversity!" one interviewee from the WBCSD said. "So we started changing the thinking, because biodiversity is a term business struggles with ... it doesn't mean anything to business, really." As a tool developer noted: "Biodiversity is such a terribly complex word, concept, [and] issue to deal with ... it's a very difficult concept to bring into, and strategize with, and think about. And ecosystems services provide a much easier way of bringing biodiversity into human decision-making."

Lohmann (2009) suggests that such a changing of the subject in pursuit of economically or market-oriented accounting practices should not be overly surprising. When "accounting practices required for new markets encounter complexities, uncertainties, nonlinearities and indeterminacies that they cannot immediately accommodate" (like biodiversity in this case), the solution is for firms, businesses, or accountants to "actively rework their objects ... to try to make them more 'passive' and tractable to the agencies of calculation" (503). The challenge of the systemic nature of biodiversity loss, the difficulty of individuating that risk, is to transform the problem away from biodiversity and toward ecosystem services.

Biodiversity is not totally excluded from the model. A note at the bottom of the excel sheet reads: "Biodiversity – the variability among living organisms within species and populations, between species, and between ecosystems – is not listed as an ecosystem service because it is not an ecosystem service in itself but rather provides the foundation for ecosystem services." As previous chapters have shown, current academic debates demonstrate that biodiversity is not necessarily implicit in ecosystem services. The extent of their overlap depends on which ecosystem service is being accounted for.[10]

Participant and "road tester" Mondi's experiences of the CESR tool demonstrate very clearly the uncertain place of biological diversity in the tool's methodology. Mondi is a leading international paper and packaging group operating in 35 countries, with 327 000 hectares of plantation forests in South Africa alone. The company's risk assessment results are provided as helpful illustration in a CESR publication (CESR 2008). This information shows that Mondi's key operational risks are related to water-based ecosystem services because their plantations depend heavily on freshwater. However, in exploring company impacts, it is notable that Mondi does not mention anything about their impact on biological diversity, in spite of wide recognition that plantation forests have impacts on biological diversity, often causing drastic changes to ecosystem composition (Brockerhoff et al. 2008). On its own, however, biological diversity does not pose a risk to Mondi's operations.

This is only one example, but it demonstrates well how this form of venture ecology focuses on identifying and understanding the narrow slice of ecosystems needed to reproduce their businesses. Returning to the example of Mondi's use of the CESR tool, the central issue becomes what kind of ecosystems and species are necessary for plantation forestry, to ensure a steady flow of paper, not what kind of ecosystems are necessary for the livelihoods of local communities, or even for ecosystem functioning. While the CESR might reveal nature's enterprising nature, it reflects a problematic conservation ethos: it calls on capital to "account for nature" but the bodies that come to matter are those that are deemed most necessary to sustain the corporation. Venture ecology, then, is not simply making visible the conditions of production, but is actively reworking the subject of biodiversity conservation, a shifting subject that will depend on the kind of firm and its needs.

The future of biodiversity business risks: Ecosystem valuation to natural capital and beyond!

While providing direction for companies to identify general risks and dependences, the CESR faces serious limitations in terms of application within the firm. As one WBCSD interviewee reported:

> Once you have ... done your impacts, there is something you can do about it. But if you are going to change the way businesses operate you need to quantify these liabilities, these externalities, and feed it into project management or forecasting, income and expenditure accounts, and your financial planning, everything businesses do. If you don't have a number associated with it, it doesn't fit in.

The demand from firms and financial institutions is for more streamlined, standardized, reliable, and cost-effective assessment tools that can produce the necessary data for corporate decisions. Managers need to know, for example, how changing land cover affects aquifer recharge, or how water availability will be affected by other changes in the structure of an ecosystem. And they need this to be tied to impacts on revenue or costs, not just in a vague hand-wavy fashion, but in specific numerical form. Just as Lohmann (2010, 229) describes in relation to the explosion of derivatives, more precise quantitative tools and techniques are needed in order to make unknowns or risks material, or as Lohmann describes, "sliceable, diceable, sellable, buyable," which, in the explosion of derivatives, was the job of quantitative experts and computer modeling.

More quantitative tools have emerged to fill these gaps, including a phase two version of the CESR created by the same collaboration between the WBCSD and WRI. This tool, the Corporate Ecosystem Valuation tool (CEV) aims to help firms turn their ecosystem impacts and dependencies into "a single (and influential) metric – money" (WBCSD n.d., 12). In early 2015, the Business Social Responsibility Network (BSR 2015) outlined 14 different valuation tools (on a list of over 120), with several targeted for corporate or financial firm use. This includes the CEV but also the "Environmental Profit and Loss Methodology," founded by none other than Sir Richard Branson,[11] and the "Total Impact Measurement and Management" tool, created by PriceWaterhouseCoopers.[12] Also included in this list of tools are Geographic Information Systems (GIS)-based spatial tools that model ecosystem service provision under different scenarios, including InVEST (discussed in the chapter 4), as well as the ARIES (Artificial Intelligence for Ecosystem Services) tool produced by Robert Costanza. These spatial tools are more detailed and quantitative – more capable of showing how a firm would be impacted due to specific land use changes, for example – but are incredibly time consuming and data intensive (Bagstad et al. 2013). There is also a growing movement around what is called natural capital accounting, with some overlap with the BBRs and ecosystem service accounting tools (see Spurgeon (2014) for a recent list of natural capital accounting tools). Natural capital accounting widens the scope and focus by including not only the work of biological diversity or ecosystems, but also what are called "natural assets" non-renewables, including water.[13] In the decade since the first reports outlining "biodiversity loss as material risk" (Table 5.1, above), this area has clearly gained steam – perhaps more than any other aspect of enterprising nature.

Conclusion

Biodiversity, recall chapter 2, becomes a global crisis by forging together ethical, biopolitical, national, and economic development interests: we must save biodiversity because the loss of a rivet imperils us all; we must save biodiversity because it is in the national interest to find new strains of crops; we must save biodiversity because we don't know which species might be the "green gold" for burgeoning biotechnology industries. In this present chapter, we see the forging of more specific calculative "risk" devices that aim to suture biodiversity loss and ecological change to the bottom line of firms, to help firms identify, mitigate, and manage risks from these changes. Conservationists are involved in the production of these tools because they hope that this suturing will lead firms to be involved in the conservation of nature, perhaps even the reproduction of social life on this planet.

These tools, I demonstrate, do not just report on the world; they reformat the world. They are constitutive and have implications, many of which are still unknown. The IBAT tool produces new spaces of investment and disinvestment, identifying the "risky" biodiverse spaces that firms should avoid. This simultaneously produces what we might call "terra bio-nullius," lands not empty of people, but rather of biodiversity of significance (at least of significance to large, multinational conservation organizations) and therefore ripe for development. The IBAT produces a smoother space for investment, helping firms avoid political risks, while at the same time sheltering some areas from developments. The CESR tool attempts to make biodiversity relevant to the operational side of corporations, encouraging firms to invest in a broader suite of less visible ecological inputs important to their production process. It solves the problem of the diffuse, systemic nature of biodiversity loss by re-orienting the problem in relation to ecosystem services, turning away from biodiversity. This is not a new process. Neil Smith, in *Uneven Development* (1984), clearly discusses how nature is produced in a capitalist mode of production. What is different here, however, is that the production of nature in the service of shareholders is put forward as a conservation strategy. The promise of a future economic system where commerce and investments can account for and conserve biological diversity is strongly desired by many in the conservation community. However, to reach this promised land, I suggest (along with Rajan 2006, Prudham 2003, Smith 1984), new kinds of capitalist natures are being produced. The work of WRI and IUCN is not simply helping businesses understand their dependence on ecosystems but also encouraging capital to invest in particular bodies and parts of nature that will maintain corporate viability.

These tools open windows into different aspects of venture ecology. IBAT shows us how conservation organizations (and perhaps ecology more generally) are finding a new niche for themselves as providers of information about biodiversity risk to firms, and creating new areas of expertise that can embed these tools in corporate decision-making. Their knowledge is becoming more valuable to firms, providing new sources of revenue; data collection becomes a "venture." Ironically, the data provided by IBAT is valuable because of the hard work by grassroots and social movements who oppose developments, creating risks to social license and permit delays. In these alignments of conservation data, ecosystem service risk assessments, and capital-maximizing firms, we are seeing, as in genomics, an implosion of "enterprises of scientific fact production with those of capitalist value generation" (Rajan 2006, 114) and an implosion of "the valuation of life with the valuations of the market" (136). Yet, unlike in genomics, these implosions are not entirely about the production of new saleable and hopefully profitable commodities. Biodiversity as risk is not a new accumulation strategy (to spin off Neil Smith's 2007 article), but rather an approach trying to produce a frictionless, riskless world for corporate transactions. It is about securing the conditions for capitalist reproduction.

While it is clear that business demands new kinds of formats and forms of environmental data to identify the specific risks to their operations and return, it is not clear if business itself is changing significantly because of these new formats and forms. In other words, while venture science is creating new knowledges and, as such, new representations for life on earth tied to the bottom line, it is unclear whether this is a strategy that actually changes the way that business does business. On the individual firm or financial institution level, there are serious questions as to how conservation organizations would even know what firms are actually doing to implement risk assessment. As one UNEP FI interviewee noted, most companies do not want to release the data on how they might actually be implementing policies on biodiversity and ecosystems. As such, this person said, it is difficult to "move from engaging with companies based on how they say they deal with it on an organizational level to actually assess them on how they implement this in practice." Business biodiversity risks are a conservation strategy that is going to be difficult to assess, an ironic situation when conservationists are increasingly obsessed with return on conservation investments.

The 2015 WEF Global Risks review suggests that little has changed since biodiversity made its Davos debut (see Figure 5.2). Biodiversity loss and ecosystem collapse are charted at a 4.5 out of 7 in terms of likelihood (with 7 being very likely to occur) at a severity of just under 5 out of 7 (with 7 being massive and devastating outcome), and so are considered one of the top 10 global risks in terms of impact.[14] The title of the section of the report on "environmental risks" says it all: "High concern, little progress." While risks are increasingly well-known by elites and by firms, "governments and businesses remain woefully underprepared." The short-term vision of both firms and states prevents them from addressing these long-term, systemic risks.

With the focus on minimizing the risks to and shoring up the well-being of corporate entities (as do the CESR and IBAT tools), there is also a chance that such conservation strategies will be in tension with the important move conservation groups are making toward socially just, human rights-based approaches to conservation, where the kind of nature that comes to matter for conservation is decided based not on hotspots or ecosystem service mapping, but on the entangled relations that many people have with nonhumans, with biological diversity. Sullivan (2013) suggests that the financialization of biodiversity

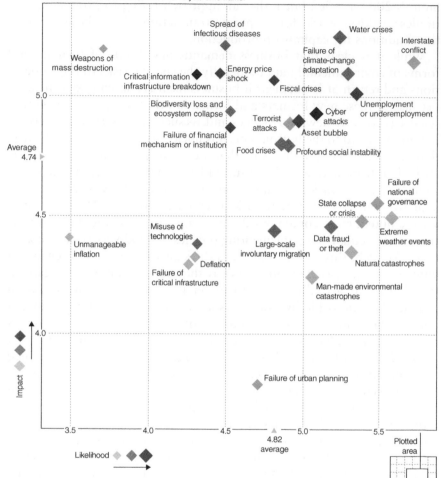

Figure 5.2 Biodiversity in the global risk landscape (2015). © World Economic Forum 2015 (by permission).

conservation is yet another departure *away* from this engagement. It effectively shifts "responsibility for global environmental outcomes into the incentivising control of investment finance" (2013, 204). Is it possible to have community-oriented, socially just, and also heavily financialized biodiversity conservation practices?

Further, all of the exclusions and uncertainties outlined in the chapter 4 related to InVEST apply here. Modeling and valuing the biophysical impacts of biodiversity loss and ecosystem changes with any precision is incredibly difficult; the ability to pinpoint the moment of drastic ecosystem function change is impossible for ecologists, or at least impossible to predict with any accuracy. The non-linearity of ecosystem change does not yet – and may never – fit into probabilistic underwriting model, especially when simple, simple, simple rules the roost. To be useable, firms (and governments) do not want complex, academic tools that require PhDs in ecology (like InVEST). How will these simplifications materialize in ecological "investments" that conservationists are hoping these strategies will yield? There are risks from prioritizing corporate accounting as conservation strategy that conservation organizations promoting this approach have yet to seriously address.

Notes

1   This chapter draws from interviews conducted with representatives from the International Finance Corporation, the World Resources Institute, Birdlife International, World Business Council for Sustainable Development, the United Nations Environment Programme's Financial Initiative, the International Union for the Conservation of Nature, and BC Hydro.

2   See also Robertson (2006), Smith (1984), Prudham (2003).

3   According to WEF (2010b), biodiversity loss–related reputational risks exist when firms' activities raise the ire of environmental groups and companies' public image is damaged. Regulatory risks are related to legal penalties that companies might face with the emergence of new government policies such as taxes and moratoria on natural resource extraction. Market and product risks arise when customers switch to suppliers that offer products with lower ecosystem impacts or when governments implement new sustainable procurement policies. Financial risks emerge when banks or financial institutions implement more rigorous lending requirements for corporate loans – requirements that take into account biodiversity and ecosystem risks.

4   In 2004, the same year as its publication of the critical first corporate BBRs report, ISIS merged with F&C Asset Management to become the UK's fourth largest fund manager, with over €100 billion under management.

5   In German: Verein für Umweltmanagement und Nachhaltigkeit in Finanzinstituten e.V. (VfU).

6   Underwriting is the process of evaluating the risk and exposure of clients and as such is an important process through which risks become tangible or material. If an insurer decides that the risks from something – say in this case changes in biodiversity and ecosystems – are too high for the insurer to absorb costs, insurance may be denied, or else constraints or exclusions be placed on the insurance contract.

7   See http://www.eiris.org/files/research%20publications/Biodiversity2010.pdf (last accessed April 25, 2016).

8   The BSR is a network of over 250 member companies that aims to develop "sustainable business strategies and solutions through consulting, research

and cross sector collaboration" (BSR 2013). The Ecosystem Service Working Group researches and reports on the "emerging risks and opportunities associated with corporate reliance on, impact to, and revenue opportunities from ecosystem services and environmental markets" (BSR 2010).

9   The video can be seen at http://www.desmogblog.com/conservation-international-greenwashing-scandal-exposed-undercover-journalists (last accessed February 9, 2015).

10 The difference between biodiversity and ecosystem services is often demonstrated with the example of carbon sequestration: if one wants to maximize the ecosystem service of carbon sequestration, this could be achieved with a plantation, not necessarily a biodiverse forest that provides multiple services to a wide array of beneficiaries.

11 See http://bteam.org/team (last accessed February 24, 2016).

12 See http://www.pwc.com/gx/en/sustainability/publications/total-impact-measurement-management/index.jhtml (last accessed February 24, 2016).

13 The definition of natural capital is usually something along these lines: "The finite stock of natural assets (air, water, land, habitats) from which goods and services flow to benefit society and the economy. It is made up of ecosystems (providing renewable resources and services), and non-renewable deposits of fossil fuels and minerals " (Spurgeon 2014, 5).

14 The WEF changed their methodology of the assessment from 2010, so it is difficult to make direct comparisons.

6

Biodiversity Finance and the Search for Patient Capital

In 2008–09, in the early, dark days of the global financial crisis, I attended two linked conferences, both called "Biodiversity and Ecosystem Finance" – one in London, the other in New York. Organized by a private London-based company called Green Power Conferences, these small, elite events sought to "bring together the corporate and finance communities with the biodiversity and ecosystem industry to drive this critical early stage market forward." In spite of the financial crisis, the participants were confident, even celebratory. Ten years ago, explained Michael Kelly, chair of the London conference and head of corporate social responsibility for the massive professional services company KPMG-UK, such a conference would have been impossible. In his view, the conference itself was a major achievement. But the conferences were also just the beginning, participants thought, and they shared a decisive belief that once the "blip" of the financial crisis passed, market opportunities surrounding biodiversity and ecosystem finance would, as one person remarked in New York, "grow and grow and grow." The number of attendees was small – 50 in London, 35 in New York – but participants displayed their optimism by blaming the financial crisis for the low turnout.

At the events, participants shared and bolstered an enthusiastic vision of a future world where biodiversity and ecosystem conservation were tools of (rather than barriers to) economic growth and capital accumulation. The chair of the New York meeting was Ricardo Bayon, a key proponent and architect of market-based environmental policies and founder of EKO Asset Management Partners, a company that

Enterprising Nature: Economics, Markets, and Finance in Global Biodiversity Politics,
First Edition. Jessica Dempsey.
© 2016 John Wiley & Sons, Ltd. Published 2016 by John Wiley & Sons, Ltd.

merged in 2015 with Wolfensohn Fund Management to form Encourage Capital. At the time of the conference, EKO was calling itself a "specialized investment firm focused on discovering and monetizing unrealized or unrecognized environmental assets," a "'merchant bank' for the world of environmental markets."[1] Materials from EKO highlighted the promise of the bioeconomy, their website telling us that a "fundamental economic transformation" is underway: "nature and nature's services are going from being essentially unappreciated and undervalued" to being seen as "central to long-term economic health." For the most part, conference participants shared EKO's belief in the economic opportunities latent in biodiversity and ecosystems. Under the right conditions, participants believed, the world's living aspects could be harnessed and their self-generating capacities could become a source of wealth.

But participants' optimism was not the only story at these Green Power events. At the conferences, I was able to observe these individuals – many of whom are key promoters of ideas about the economic potential of biodiversity – in their "natural habitat," where they spoke freely about the practical challenges associated with bringing this new type of economy into being. By attending the London and New York events, I was able to study elites in the biodiversity industry in the type of free discussion that is often inaccessible to social science researchers.[2] What I witnessed was a discourse that – on the one hand – celebrated the promise of the bioeconomy and suggested that economic opportunities were forthcoming and significant while – on the other hand – enumerated barriers to the creation of biodiversity- and ecosystem-related commodities and markets.

The take-home message from these conferences was simple: creating a profitable enterprising nature was a process fraught with uncertainties and challenges. International professionals, conservation actors, and government representatives struggled to figure out exactly what they needed to do to make biodiversity a part of market vocabulary and – even better – an economic actor that could save its own life. In a presentation on how to market forest carbon investments to institutions, a senior advisor at a boutique UK-based investment firm pushed participants to clearly define the commodity. "Investors buy a certificate on note representing ecosystem services," read his first PowerPoint slide. "*What exactly are they buying?*" The questions proliferated: "What rights are associated with these notes? Is it income? Is it asset backed, e.g. land?" Say "a pension fund resells an ecosystem-based note," this speaker asked, "how do they calculate the price?" And what "is exactly being 'transferred at sale'?" With each question, the outline of the commodity grew blurrier.

At the conferences, actors discussed how to create or expand biodiversity markets, attempting to make new aspects or parts of nature into sites of accumulation, into – as Sian Sullivan (2013, 210) describes – "reified, exchangeable, and financialized commodity forms."[3] As one participant said excitedly, without a hint of remorse, in a breakout group, "What we are talking about is the privatization and commodification of nature!"

My approach to these ideas, here, is to bring readers into the "natural habitat" of different "species" of actors involved in the generation of ideas and practices related to biodiversity markets. I explain the history and state of biodiversity markets through the language of conference participants to show the tensions in their vision; this approach, I believe, provides a useful entry point into understanding "Nature Inc." (Büscher et al. 2014). Conference participants espoused the ever-hopeful discourse of the bioeconomic promise while repeatedly stumbling over the abstractions that would make the bioeconomy operational. By staying close to these conferences, to the presentations and projects discussed, I draw out the challenges of biodiversity market-making. These initiatives – in contrast to the lightning-fast liquidity of current capitalist relations – are slow and lumbering. The biodiversity and ecosystem "industry" remains as a halting endeavor, a set of partial attempts to create new facts of life that will facilitate commodification and allow biodiversity to fit into financial models and market logic. While conference participants often celebrated biodiversity market-making processes as a source of significant opportunity, these actors also described the markets in terms of what they lack: a convincing language, a defined unit of measurement or commodity, scientific data, financial understanding and data, and regulation.

The State of the "Markets"

At the New York conference, one US-based eco-entrepreneur asked participants to linger awhile on the words of Galileo:

> What greater stupidity can be imagined than calling jewels, silver, and gold "precious" and earth and soil "base"? People who do this ought to remember that if there were a greater scarcity of soil as of jewels or precious metals, there would not be a prince who would not spend a bushel of diamonds … to have enough soil to plant a jasmine in a little pot, or to sow an orange seed and watch it sprout, grow and produce its handsome leaves, its fragrant flowers, its fine fruit.

For this speaker, Galileo's words point to critical notions of both scarcity and value, two pillars of liberal political economic theories. At the core of Galileo's quote, according to this eco-entrepreneur, is the idea that scarcity drives value. Making reference to the widespread loss of biodiversity, he went on to say, "We are getting to that point [of scarcity] with biodiversity, and the various ecosystem services." This scarcity, he argued, "is going to drive value."[4] While there are many ways to value biodiversity, for this speaker and many gathered around the conferences, the best and most efficient approach is to price it, a process requiring markets. He said: "Markets are extremely important, because they are a way to put value on biodiversity. And I think we are going to see more and more of them." In his presentation the speaker provided a figure to demonstrate this growth: a line graph showing the size and growth of environmental markets, including sulphur dioxide/nitrous oxide, biodiversity, voluntary carbon, EU carbon, US carbon, and water quality markets. Into the future, the lines for these markets all pointed up, indicating steep exponential growth for the carbon markets, and a nice upward-sloping line for biodiversity markets.

A staff member from Ecosystem Marketplace explained the biodiversity market types in New York, dividing markets into three categories: policy driven, information driven, and cap and trade (in 2010 Ecosystem Marketplace would release its first "state of the biodiversity markets" report; see Madsen et al. (2010)).[5] In this chapter, I focus primarily on this third category of biodiversity markets – cap and trade – because it involves the creation of new commodities and, ideally (or so the argument goes), new flows of capital toward conservation. A cap-and-trade biodiversity market is a system wherein governments set a cap on development on ecosystems of a particular type but create provisions to allow impacts to these ecosystems in exchange for protection or restoration of the same type elsewhere.[6] This protection of one site in order to offset impacts at another is also called "mitigation banking."

Mitigation banking, as reported at the New York conference, was about a $3 billion industry and growing steadily, with the value drivers being infrastructure development (including real estate) and energy production and distribution.[7] In mitigation banking, market value – and the resulting protection of biodiversity – is driven by development. Without ecosystem change somewhere, there can be no market. Around the same time of these conferences, one journalist noted, without irony, that while the recession might be good for the environment (e.g. less fossil fuel–energy infrastructure), it was bad for the mitigation banks that make a living off saving species (Kenny 2009).

Markets like wetland and species banking are beginning to put a price on new things; these prices are determined by profitability in the development-related market. The eco-entrepreneur noted how one endangered species, the Delhi Sands flower-loving fly in California – the first fly on the federal endangered species list – stopped development in the San Bernardino

County (one of fastest growing counties in the US). One mitigation bank operated by Vulcan Materials Company sold credits to the developer for $100000 and $150000 an acre for fly habitat (Campbell 2006). The same eco-entrepreneur laid out the dilemma:

> The question this poses, and this is a broader question for the biodiversity markets, [is] to what extent does this really tell us how much biodiversity is worth, and how much should it be worth. How much should this fly ... be worth? ... How much should that fly be worth in our society, and how do we decide that? And how do we put a price? And who puts the price? And who pays that price? And what are we giving up in return for that? These are the questions biodiversity markets need to answer ... that society needs to answer going forward.

In the San Bernardino County case, the fly was worth however much a developer could pay to meet its return objectives. So the value of the fly and its habitat was determined by the developer's profit, which was influenced by the housing market: the amount the developer was able to sell houses for, the cost of land, plus all the regulatory costs, including development permits. Through complex negotiations involving many state and non-state actors, a mitigation bank could put an economic value on biodiversity. The eco-entrepreneur's quote reveals, too, that these markets put prices on species, but that these prices were, as with all prices, fraught indicators of social value. Conference participants did not take up this tension between a market approach to value discovery and other, perhaps more democratic means to determining the social value of a fly; neither did they consider responsibility for the ongoing loss of fly habitat.

Hovering over the conferences, too, was the question of whether this kind of valuation actually worked to conserve or protect biodiversity. At the London event, one audience member posed a key question to a government panel: "From a policy perspective, should we be looking at offsetting or avoidance?" He continued, "Offsetting implies loss somewhere else." And: "The way things are going, do we have time for this?" He wanted to know if – given the rapid deterioration of the global environment – the loss of portions of any biodiversity-rich ecosystems was justifiable. The responses of members of a government panel essentially concluded that any offsetting needed to be done carefully, to ensure that it did not undermine avoiding impacts. The question, though, of whether biodiversity markets result, or might result, in a net protection of ecosystems and biodiversity is a very real one.

The scientific data necessary to answer this question is lacking; even wetland banking, the most advanced banking system in the world, has not been subjected to a systematic assessment (Robertson and Hayden 2008).

However, some evidence suggests compensation and offsetting are taking priority over the other aspects of most banking schemes, which require proponents to avoid and minimize impacts prior to offsetting (Hough and Robertson 2009; Clare et al. 2011). Some scholars argue that offsetting programs such as wetland banking might be working perversely, in some cases, as incentives that support ecologically problematic developments.[8]

The savvy response of the government official – that offsetting needed to be done "carefully" – is emblematic of the kind of contradictory discourse at the conferences, which highlighted, simultaneously, and in this case very succinctly, both possibility and practical obstacles. It is safe to say, though, that at these two events, the question of the conservation effectiveness of biodiversity market-making was a topic secondary to the main flow of discussion, which centered on the specifics of how to make new commodities and how to make private capital flow toward conservation. Broader ecological concerns were often bracketed, if they were discussed at all.

Thus, at the Green Power events participants celebrated the promise of new commodities while also addressing, often as practitioners, the practical challenges associated with bringing these commodities into being – or, with existing markets, as in the case of wetland banking, they focused on how to bring more capital into them. While in this chapter I explore the idea of biodiversity market-making in practitioners' own words, this contradictory discourse needs to first be situated within broader global economic and environmental policy contexts; the tensions within the discourse make sense when understood within the moments of upheaval within which the conferences are situated. The Green Power events took place at the beginning of two international crises: the global financial crisis and the crisis in the international carbon market. Yet, as we will see the conversations at the conferences was strangely selective, overlooking some of the most pressing environmental-economic issues of the day in favour of general optimism and the nitty-gritty practical talk of who could be selling what.

Illiquid natures and the global financial crisis

Both conferences, in London and New York, bore the indelible mark of the financial crisis, a crisis precipitated in the under-regulation of new financial products that led even the most ardent free market supporters such as Alan Greenspan to express some lost faith in the wonders of the market. One would think that the financial crisis might lead to similar questioning of market-based environmental policies. Overall, many market advocates I observed at the London and New York conferences demonstrated unmitigated and unabashed confidence in the approach. As one investment banker in New York stated:

> This is a market that is very Churchillian. What would Churchill say in this market? I think he would say we are at the end of the beginning, we are hitting the middle, there are going to be losses before wins. The winners are going to be those who have some staying power – what that means is conserving cash, being around, surviving, surviving is winning now. When the market comes around, the money is going to go in this direction [i.e. the direction of biodiversity market-making]. If you position yourself for low risk low return profile, you are most likely to have more institutional investors in the near term. We are seeing firefly investments now, but are they fireflies that will die, or be the start of the trend?

This remark, in some ways, encapsulates the general tone of the conference. While participants in these events seemed to truly believe that there is "no alternative" to a fully enterprised and profitable nature, they also displayed uncertainty. Will these approaches be robust, life-giving fireflies, or will they "die out"?

When these markets are viewed within the broader landscape of global capitalism, the light from any fireflies looked very small and very dim indeed in 2009. Recognizing the need to compete for ever-scarcer resources in a contracting economy, in the midst of a "credit crunch," event participants perhaps felt compelled to speak in laudatory, promotional terms about the promise of the commodities they sought to create. They described the crisis as a lull in an otherwise upward trend. They sought to show present biodiversity-related commodities as a future certainty, as part of a market that had all but arrived.

The crisis in the carbon market

In his opening presentation at the New York event, the eco-entrepreneur followed up his slide showing the exponential growth of environmental markets by comparing the biodiversity market-making moment to carbon markets: "Markets are being created, prices for biodiversity are beginning to appear. But it is still very much the beginning. This isn't carbon, yet." He added: "And it probably never will be … because biodiversity is very, very different from carbon." Biodiversity, this same speaker said, does not have the "natural commodity" form that carbon does, which is the tonne. In the carbon market, all six greenhouse gases are expressed in terms of their carbon dioxide (CO_2) equivalents, which means that there is a common unit for exchange. Again and again, Green

Power conference participants expressed a kind of envy of the carbon market, and they often explained their challenges in relation to the markets in carbon. Participants viewed carbon as a commodity that is easy to measure, trade, and market.

Conversely, there is no one measure for biodiversity, and thus no singular biodiversity market. In fact, there is no biodiversity market at all. Existing markets are not for the total aggregate of biological diversity; rather, they relate to specific aspects of biodiversity such as endangered species or wetlands or are tied to specific ecosystem services such as the carbon sequestration of tropical forests. Later in the day, another speaker pinned the fragmentation of biodiversity markets on "the very nature of biodiversity." This person went on to say that "in carbon you have a natural-born commodity, in biodiversity you have the world's greatest anti-commodity where its value is in its diversity." Many conference participants noted the lack of quantitative indicators and measures as key limitations for biodiversity markets. As an economist from the International Union for Conservation of Nature (IUCN) said:

> In the case of biodiversity we don't yet have that coherence, that consistency, that consensus on what indicators to use [to measure changes in biological diversity] ... If we don't have targets and indicators in quantitative terms that are comparable ... across sectors, across countries, across instruments, then we are not going to be able to make headway.

He illustrated this point with a slide comparing the various ways that biodiversity and ecosystems are measured with techniques found in the climate change world. Biodiversity can be measured 11 different ways – including species richness, habitat hectares, protected area coverage, natural habitat fragmentation, inter-species diversity, and ecosystem services. No single unit like a "CO_2 equivalent" exists for biodiversity.

But is carbon really a "natural" commodity, as actors at the Green Power conferences claimed? At the time of these events, major criticisms of two of the world's biggest and best-established carbon markets – the European Trading Scheme (ETS) and the Clean Development Mechanism (CDM) – were circulating widely. There was already widespread recognition that the ETS was fraught with problems due to a cap set too low and an over-allocation of permits. This led the UK Environmental Audit Committee to claim that phase one of the ETS was likely to be "ineffective in driving down emissions" (cited in Gilbertson and Reyes 2009, 34), the main purpose of the market. A more scathing review of the EU ETS suggested that, among other problems, the trading scheme might actually be rewarding companies that emit disproportionate amounts of

CO_2 (Dorsey 2007). By 2010 the CDM was also already understood as underperforming in one of its core aims: to develop low carbon energy in developing countries, with many of the reductions in "low-hanging fruit" that failed to establish a trajectory toward a low-carbon energy infrastructure in the Global South (Wara 2007). In addition to these problems, the market was seen to lack on-the-ground enforcement (Creagh 2009).

By the time of these Green Power events, debate over whether the carbon market was failing at its stated purpose was public and widespread, but actors at the events scarcely mentioned these issues. Instead, the carbon market was held up as a model by participants, such as by the speaker from IUCN. This continued bracketing of the carbon market crisis (and other major challenges and contradictions) is illustrative of how this group of actors reconciled tensions in their attempts to create biodiversity markets. At the Green Power conferences, participants created shared blind spots in order to maintain the optimism necessary to promote these new markets and pursue investment. Performances of optimism and hype are crucial to new markets and products, whether it is a new gold mine in the Kalimantan (Tsing 2005) or the possibility of a new drug (Rajan 2006), and environmental markets are no different. One does not sell a policy or product through endless loops of nuance and deep investigations of failure, but rather through dramatic performances (see also Sullivan 2013). To draw from the title of one of Tsing's chapters, one must keep up the "Economy of Appearances."

A Taxonomy of Actors and Projects

> This is a small group of people, which makes it easy to talk to each other.
> (Participant in London conference)

What was the draw of biodiversity market-making for this small but growing network of people? In the next section, I characterize six different "species" of actors at the London and New York events and explain their interest in the project (Table 6.1). These species were, by and large, white and Western. At the risk of making assumptions, I believe that, for many, their natural habitats were conference rooms, major cities, and airports. Meanwhile, however, they shared with many of their biodiversity-interested predecessors (many of whom I've described in the first five chapters of this book) a southward gaze, an interest in the ecosystems of the biodiversity-rich Global South. This was particularly the case in the London conference; the New York event had more focus on the US, especially given the already existing

Table 6.1 "Species" of actors at the Green Power conferences.

Species	Function	Types of actors	Examples of specific groups	Interest/Role
A	We deal with biodiversity through voluntary corporate social responsibility and risk management	Private banks and multinational firms	Holcim (cement manufacturer); Citibank; Rabobank	Promoting voluntary initiatives to demonstrate company responsibility and manage environmental, social, and governance risks
B	We are creating new financial products that can attract investors	Specialized investment banks; think tanks; international investment organizations	EnviroMarket; Forum for the Future; International Finance Corporation; Merrill Lynch	Sharing information on the creation of value and bringing a realistic investor perspective
C	We are creating return-generating biodiversity conservation projects	Small independent companies; NGOs	Bio Assets; Canopy Capital; Global Canopy Programme	Finding investors for the projects
D	We seek additional resources and avenues for conservation	Conservation groups	WWF; Fauna & Flora International; IUCN	Developing partnerships to generate revenue for conservation

E	We provide the technical support, information, and infrastructure new markets need	International organizations; private environmental registry companies	UNEP World Conservation Monitoring Centre; TZ1 Registry; Ecosystem Marketplace	Offering data on conservation or promoting registry services
F	We develop regulations and rules (sometimes) that aid market development	Governments	London conference only, included a director from the CBD Secretariat and senior bureaucrats from the Netherlands Environmental Assessment Agency; the German Ministry of Environment, Nature Conservation and Nuclear Safety; and the European Commission.	Developing support for policies and initiatives that will generate more financial resources for conservation

biodiversity markets in wetlands and endangered species. Overall, according to information provided by Green Power Conferences, attendees at their events are a mix heavy on banks and investors (23%) and conservation groups (21%). Other participants include consultants (12%), representatives from the oil, gas, and mining sector (8%), universities (9%), government (9%), and a smattering of other sectors. Taken alone, these numbers do not provide much insight into the intentions and activities of participants, and in this section, I try to give some shape to these numbers by categorizing participants by their key aims and, therefore, their functions in biodiversity market making. Like all typologies, these are imperfect – some actors blur into other categories or perhaps have dual or triple functions.

Species A: We deal with biodiversity through corporate social responsibility and risk management

Several of the speakers from large banks and multinational firms focused on their voluntary initiatives to manage the impacts of their industries on biodiversity. As such, the focus was often through the lens of corporate social responsibility (CSR). CSR can be considered part of a broader "information-driven" market, wherein the reputation of the company becomes part of consumer decision-making. The CSR projects presented were about selling existing commodities (e.g. cement) and risk management, not about new biodiversity commodities. At two presentations I attended, corporate representatives emphasized the voluntary initiatives their companies had undertaken to consider biodiversity loss. A representative from Holcim, one of the world's largest cement manufacturers, described the company's voluntary collaboration with IUCN to improve their operations' impact on biodiversity loss. A biodiversity specialist from Citibank presented a case study of their investment in production of a liquefied natural gas, activities for which the company demanded intensive mitigation and coral transplantation. The requirements for these compensatory measures did not arise from legislation; rather, they arose as part of the bank's standards as signatory to the Equator Principles.[9] The Citibank participant told me that when she was preparing for the conference, she looked for an example of an investment or loan in a "biodiversity market" but was not able to find one (something I will return to later). The overall purpose of these presentations was promotional, demonstrating how far these companies had come in raising biodiversity-related concerns in their operations and lending, an effort that parallels the risk-management work I described in chapter 5.

Attempts to "green" existing businesses are not the central subject of this chapter. It is worth, however, understanding the challenges that these actors face in trying to take account of biodiversity, since the failure of voluntary corporate initiatives to protect biodiversity is another example of the kind of market failure that biodiversity market-makers seek to correct by creating biodiversity-related commodities. A representative from Rabobank (one of the world's 15 largest banks, and one with connections to large biodiversity impacts through loans to agribusiness) spoke openly about the challenges the company faces in relation to reducing impacts. The representative spoke plainly about the impacts their investments have on biodiversity. Surprisingly, he began his presentation by showing an image of a massive sugar cane plantation in Brazil that caused large-scale deforestation, a development that this bank actually funded. The bank, he told us, was trying to incorporate environmental, social, and governance principles – including those that take into account biodiversity and ecosystems – into all aspects of their lending, particularly through the training of desk staff who process and approve credit applications in offices all over the world. However, he noted, the point of incorporating biodiversity or ecosystem services into their project lending criteria was not to reject applications, which the bank rarely does. Rather, members of the desk staff are encouraged to work with project proponents to change their approach, encouraging them to be more sustainable in their agricultural practices. The speaker explained that this is because those seeking loans will simply find financing elsewhere if they are rejected. He went on to say that the "problem is not the big multinationals ... the big problem is the smaller companies that have not integrated sustainability." In this way, he positioned the bank as not only a capital provider but also a teacher and leader. However, he also noted candidly that most staff received their incentive pay based on the volume of loans made; any rejection of loans due to habitat concerns would therefore be difficult.

While no one at the conference said it, the situation at this bank provides evidence to support a common sentiment among activists and some NGOs: the road to planetary ruin is paved with relatively powerless vice-presidents of sustainability. The sentiment refers to the lack of power in CSR departments and the limitations of a voluntary approach and risk management approach to stemming biodiversity loss. Even executives who "get" the problem of ecosystem loss cannot undercut the profit-making motive of their organization. In the case of the bank, its main purpose is to make loans to those who are going to pay them back, and make the bank's shareholders a return from these loans. In this way, the CSR presentations either promoted the companies involved or gave examples of the limitations of the approach and thus seemed to help solidify the belief of other actors: biodiversity must be made a source of economic value in and of itself.

Species B: We are creating new financial products that can attract investors

Species B was made up of people who focus on the creation of new business opportunities and financial products. A boutique private equity asset manager, also a former Credit Suisse executive, characterized her firm and the investments they make as focused on money-making, investor return-generating opportunities, where the environmental and social benefits are part of the economic value driver. She clearly stated that she was not talking about CSR or even risk mitigation, but rather on "making money for investors where there is a real economic value proposition." This banker was also not talking about investing in small biodiversity-friendly businesses via small loans or micro-credit, or what might be called impact investing. Her question, which she repeated over and over again to the group assembled in New York boiled down to this: What is the economic value driver in this space? And how can we ramp up biodiversity markets so that they become intelligible to mainstream investment?

Representatives from capital management firms emphasized the need for what some at the conference described as "patient capital": long-term, low-yield investment vehicles that would be attractive to investors who do not demand quick return or payback, who are comfortable with less liquidity in their investments. This patient capital provides time for biodiversity value to materialize. A speaker from the International Finance Corporation (IFC), noted that the IFC was currently developing such investment vehicles, a project in collaboration with the UK-based think tank Forum for the Future and private sector organizations like EnviroMarket. Other asset managers emphasized the need for patient capital, highlighting the need to connect the right investors – those who will be satisfied with long-term returns of say 6% per annum, over a 30-year period or more (e.g. pension funds) – with the right projects, providing those kinds of returns.

Several participants in New York and London were involved with or excited by the creation of financial products like "forest-backed bonds" and "eco-securitization." The purpose of the forest-backed bonds is to help some institutional investors shift from capitalizing forest plantations – heavily managed tree monocultures that supplied a steady and predictable return – to investing in "natural forests," which might grow a diversity of tree species and also provide all kinds of other socio-ecological benefits, particularly in the form of ecosystem services like carbon sequestration, watershed protection, and biodiversity conservation (Petley et al. 2007). These bonds propose to generate value by merging traditional asset classes with environmental markets. For example, this speaker reported, a patch of tropical forest could produce some sustainable timber sold through timber markets, as well as receiving payments for carbon sequestration and watershed protection. The creation of these bonds is by no means a foregone conclusion, as they require significant institutional infrastructure in order to function: development of this credit pool, notes an IFC report, depends on "the ability to secure long-term off-take agreements with national governments for certified timber and carbon, and with multilaterals for carbon" (Petley et al. 2007, 3).

At this time, as the same investment banker kept reiterating, investment capital is certainly not knocking down the door: "There is lots of stuff going on and lots of discussion. But when you think about it, very little is going on." In her investment models, biodiversity, carbon, and water are still valued at zero in terms of revenue. However, as she went on to explain, these models do include them in the model as "upside scenarios" that potentially have value in the future, but only for high-risk investors. Other conference participants, such as a speaker from Merrill Lynch, noted that a key challenge with the voluntary forest carbon market is that the assets are illiquid and cannot be cashed out quickly. In other words, developing financial products that require patience on the investment side is itself a major barrier to the growth and development of biodiversity markets. With their promotion of "patient capital" in an era of lightning-fast financial movement, Species B representatives appeared to be facing a major struggle to break into even the margins of the investment market.

Species C: We are creating return-generating biodiversity conservation projects

Species C was composed of project proponents seeking investment. For example, Bio Assets, a company working in carbon and biodiversity assets, aims to create value from protecting nature and, in particular, protecting a 100 000 hectare site at the mouth of the Amazon River, an area more than 11 times the size of Manhattan, which Bio Assets owns outright. Bio Assets seek financial returns on their investment in land based on private reforestation and conservation of forests.[10] At the same time, they are also are striving to generate projects that can potentially create income for local communities. In order to carry out these reforestation and conservation projects, they seek what they called "enlightened" investors and collaborators, which was partly why they attended this conference. The project aims to achieve a kind of "win-win-win": a win in biodiversity outcomes, a win in "development" outcomes, and a win in profit (the so-called triple bottom line).

The attempt to achieve a triple win, however, is rife with difficulties; a representative at the conference noted that the project faced "severe challenges," including challenges in methodology, implementation, monitoring and verification, and "community involvement," including issues of land tenure. Understanding the social relations for the project requires more study and fieldwork, but here I focus on how the company explained the value in the land and for their project. Market value emerges from bundling together a number of investments in the forests, some that are partly traditional and some that are partly based on emerging markets in environmental services (i.e. those in carbon emissions trading, watershed protection, and biodiversity). The project investments demonstrate something like a "sustainable forests plus." The Bio Assets presentation listed five revenue streams:

1   Reforestation activities, which could capture revenue in both regulated and voluntary carbon markets. Project proponents estimated that reforestation would have capital expenditure cost of $8–12 million with an internal rate of return (IRR) of 15–20%.
2   Avoided deforestation activities under the Reducing Emissions from Deforestation and forest Degradation initiative (REDD),[11] with capital expenditures of $10 million and an IRR of 30%.
3   Wood and palm reforestation activities (including tradable products, but also potentially certified carbon emission reduction credits), with capital expenditure costs of $6–9 million and an estimated IRR of 20–25%.
4   Watershed protection services with capital expenditure costs of $5 million and an IRR of 40%.
5   Wild biodiversity conservation with capital expenditure costs of $5 million and an IRR of 40%.

With these "stacked" revenues, the Bio Assets project proponents believed they could generate profit for their investors, achieve conservation and restoration, as well as provide local people with livelihoods. It is impossible to analyze the plausibility of this deal with this information alone; my purpose here is simply to show how project proponents understood how value could be generated for the investors in their Amazon site.

Another project seeking investment was one created by Canopy Capital, a private company that is an offshoot of the UK-based NGO Global Canopy Programme. Canopy Capital is owned 20% by Global Canopy Programme and 80% by investors. In March 2008, Canopy bought the rights to measure and then value the ecosystem services provided by a large forest in Guyana, the Iwokrama forest, for a period of five years.

In exchange, Canopy would make a guaranteed yearly payment to the body that manages the reserve. At the time of the presentation, Canopy Capital was exploring various approaches to fund their payments to the Iwokrama forest; they had ideas similar to those of forest-backed bonds of Species B that would offer returns based on the sale of ecosystem services, what they called the "eco-utilities" provided by the forest.

Yet by 2012, Canopy Capital had paid £500 million for the rights to Iwokrama's ecosystem services but had failed to sell a single bond, its bankers concluding that "as yet, there is nothing to sell" (Pearce 2012). That same year, a Guyanese newspaper reported that the deal had collapsed, and that Iwokrama would stop receiving payments from the British investment group (Stabroek News 2012). It turns out that the investors lost their shirts on the deal. From a distance, it looks as though the project collapsed because of difficulties establishing agreements with national and community actors over control and flows of ecosystem services from the forest, but also because of the broader non-emergence of a forest carbon market.[12]

The Iwokrama example suggests that Species C – those making marketable projects and assets related to biodiversity conservation – face challenges in dealing with local people who are already using the land, and who might stand to benefit unevenly or even lose from the offset project. They also continue to face difficulties with national governments themselves struggling with establishing legal and policy frameworks that could make more investors feel comfortable with so-called biodiversity markets, or even at the intersection of biodiversity and carbon markets.

Species D: We seek additional resources and avenues for conservation

For Species D, composed of mostly conservation organizations, most of the primary focus was on accessing more resources for conservation, with many noting the increasing difficulty of obtaining them from public sources. A representative from the Royal Society for the Protection of Birds, a UK conservation group, stated the problem clearly: "In 2002 we asked a number of very clever academics precisely how much money they would need to manage a worldwide network of protected areas on land and sea. They said 45 billion. I know we are not going to get this from the public sector. We have to think about other alternatives and opportunities." Most major NGOs like the World Wildlife Fund and The Nature Conservancy have programs in conservation finance. For conservation organizations, the sources of conservation funds and opportunities to increase these funds are critical issues, widely discussed.

Some conservation finance is focused on small-scale projects that support the development of conservation-oriented businesses. One conference participant described a project that gave funds and technical assistance to biodiversity conservation-based ecotourism businesses in Eastern Europe. Another presentation focused on work by WWF, the development-focused NGO CARE, and the International Institute for Environment and Development (IIED): a project that exists to create what they call "payments for watershed services" (PWS) in five countries (Guatemala, Peru, Indonesia, the Philippines, and Tanzania). These two were funded by donors, the Dutch and Danish governments, but they aimed to establish self-sustaining businesses or contracts (in the case of the ecosystem services) that they hoped would persist long after the donors withdraw. By 2010, a total of 650 farmers had become "sellers" in the initiative, being paid for the efforts to improve land (Lopa et al. 2012). Some initial assessments of the Tanzanian PWS scheme identify certain ecological benefits (see, for example, weADAPT 2012), but also suggest that the program is having trouble moving beyond donor funding to create permanent mechanisms through which local people can be paid for the watershed services they manage (Lopa et al. 2012).

A key development in the past few years is that international NGOs are allying themselves with financial institutions not only to influence investments and projects in pursuit of "best practices" but also to develop new avenues of value and financial investment. One example presented at the conference was the partnership of the NGO Fauna & Flora International (FFI) with Macquarie (an Australian bank) to protect six forests at risk from deforestation, in a REDD (Reducing Emissions from Deforestation and forest Degradation) project. As presented, FFI was to work with local communities and governments to manage the preservation and implement the projects, and Macquarie was to provide the capital and financial services and ensure compliance with carbon standards. Macquarie would then sell the carbon credits internationally; these sales would be the main source of income for the project. This partnership rolled into a new company, Biocarbon; in 2011 Biocarbon became backed by three investors – Macquarie, the International Finance Corporation and a private equity fund, Global Forest Partners LP.

In short, conservation organizations have created several partnerships with the goal of establishing self-sustaining enterprises in which biodiversity's life-generating capacities, as locally managed, will produce value. Some appear to take off, like Biocarbon (although, interestingly, FFI is no longer involved); others face difficulties in becoming financially self-sustaining.

Species E: We provide the technical support, information, and infrastructure new markets need

Before any of these new financial products can exist, a large infrastructure must be put in place, a set of complex socio-economic arrangements that can allow the market to function. At these conferences, Species E was those providing and managing information, including ecological information, market information, and, ideally, a way to bring these two types of information together. For example, what are the ecosystem services that the Iwokrama forest provides? What is their economic value? How will this value be monitored over time? How will trades in ecosystem services certificates or bonds be tracked and priced? What will the legal contracts for sale and service look like? The creation of new markets clearly requires technical support, information, and infrastructure.

Focusing on ecological data and support, a speaker from UNEP's World Conservation Monitoring Centre (UNEP-WCMC) noted that biodiversity data is a complex field and that it was not easy for the corporate and financial sectors to get the information they need to make good decisions. To fill this niche, UNEP-WCMC recently collaborated with BirdLife International and Conservation International to create the Integrated Biodiversity Assessment Tool (IBAT) (see chapter 5), which provides companies with up-to-date biodiversity data to support their decisions (and, ideally, to move companies' operations away from important biodiversity spaces).

Another example of an organization meeting the technical, informational demands of market creation is the TZ1 Registry; an executive from this business spoke at the New York conference. TZ1 described itself as an "environmental registry service," providing what we might think of as paperwork services, ones that verify that things are being bought and sold and by whom. They manage the "life cycle" of trades that go through the Voluntary Carbon Standard (VCS), for example, from issuance and allocation all the way through to retirement. The representative spoke of the need for emerging markets such as biodiversity to have the right information infrastructure, such as theirs, infrastructure that would enhance credibility and transparency of the market and prevent double selling. In the manner of a land registry tracking the exchange of properties in real estate markets, institutions like TZ1 are needed to manage the information and transparency of a new market. At the same time, being part of the market infrastructure can be highly profitable. In 2009, financial information services company Markit purchased TZ1 for over $37 million. Another key actor in this area of market information and infrastructure is Ecosystem Marketplace – a US-based non-profit organization that actively participated in the New York conference.

Ecosystem Marketplace is an online information source for markets and payment schemes for ecosystem services (and source of the "state of the biodiversity markets" report, described above; see Madsen, et al. (2010)); they characterize themselves as the "Bloomberg for the emerging ecosystem markets." One of their most recent projects is a website called speciesbanking.com, an information clearinghouse that aims to increase the transparency of markets related to biodiversity offsetting, compensation, and banking by providing information on prices, locations, and so on.

Species F: We develop regulations and rules (sometimes) that aid market development

Governments' participation in these conferences was limited and marginal at best. In the London conference, government representatives were jammed onto one panel and the New York session had no government participation at all. When conference participants asked about the possibilities for a Kyoto-style instrument for biodiversity under the CBD, the bureaucrats gently demurred that this was a political issue, one for individual Parties (signatory governments to UN conventions like the CBD) or governments to decide upon (see chapter 7). Panelists in several sessions highlighted challenges for government action; one issue that came up regularly was that any new standards for commodity imports can face challenge from the World Trade Organization.

However, while there was not active participation from governments, talk of governments and regulations was everywhere. In particular, participants emphasized the need for governmental intervention to create the conditions within which their initiatives could flourish, in particular the political will (as mentioned at the start of this chapter) that could bring in new laws and policies. In the next section, I discuss this point in depth, as part of a broader explanation of the challenges all these "species" of actors face in the project of biodiversity market-making.

Making Markets: Challenges

As the above "taxonomy" demonstrates, a variety of actors make up biodiversity and ecosystem finance, each with different motivations and functions for biodiversity market-making. The above section also introduced the kinds of elements or conditions that are necessary for market development: actors with new and hopefully return-generating projects requiring investments; firms focused on market infrastructure like registries; organizations providing ecological data and calculations; as well as financial institutions, asset managers, and of course investors willing to put their money on the line. In other words, there is a growing web of relations that aims to make markets in biodiversity and ecosystems, to create the conditions upon which diverse natures can be made "enterprising," where biodiversity can become a true source of economic value.

Yet creating these conditions is extremely difficult; complex intellectual and political gymnastics are required to move these initiatives forward on sound financial and scientific bases. Furthermore, as the above descriptions of projects showed, many initiatives develop complex infrastructure and cross-sectoral partnerships only to sometimes falter at the point at which conserved biodiversity might actually generate economic value. Why do these innovations in economic and socioecological relations so often fail to live up to their promise? Here I lay out five key challenges for emerging environmental markets, as articulated by participants of the two "Biodiversity and Ecosystem Finance" conferences.

Challenge 1: Lack of convincing language

A strangely recombinant discourse emerged at the two conferences. The keynote speaker in London, David Bellamy (one of George Monbiot's "top ten climate deniers" – see Monbiot (2009)), coined the terms "biodiversity crunch" and "Amazonian crunch," drawing together ideas of the credit crunch of the time with deforestation. (The chair of the session fixed on this language, saying, after Bellamy's talk, "Hands up anyone who isn't going to be using the term 'biodiversity crunch' in the next couple weeks, or the term 'forest crunch.'") Later that same day, someone quipped, "bankers are an endangered species," producing many chuckles and knowing glances around the room. With these catchphrases, participants seemed to effortlessly merge economic and ecological ideas. As I've just described, too, I noted growing alliances between NGOs and the financial community.

At the same time, however, there remained large gaps in language and understanding between finance and conservation. In New York, one eco-entrepreneur presciently argued that the main challenge for emerging markets in biodiversity was convincing people that the market existed and that it had lasting power. In his presentation, he said that there was a need to understand how to classify emerging biodiversity markets (such as species banks). He felt that tremendous education was needed, because very few people understood biodiversity markets. Part of the challenge is the multi-disciplinarity of these markets.[13] One US-based conservation bank consultant explained the problem in simple terms on his PowerPoint slide:

$$Science + Law + Economics = challenge.$$

The boutique investment banker present in New York noted the difficulties of terminologies in use, particularly the use of the term "bank" in reference to mitigation and conservation banks. She said that such terms were extremely confusing for those in the investment community. She went on to say: "A lot of funds, they have analysts that cycle through, one day looking at energy, another day on paper, then on retail. They go in cycles. The term 'bank' throws them for a loop. It loses time, causes confusion." She made the same point about ecosystem services, noting that investors still do not understand what the term might mean for them. She continued: "The way we explain it is to divide it into revenue and costs. Water is a potential revenue source, carbon is a potential revenue source." In other words, biodiversity must be translated into something even more like a revenue flow before financial analysts will begin to assimilate it into everyday calculations. This translation demonstrates the important role that linguistic codes play in describing and ultimately making novel financial products, and making them acceptable and widespread (see Lépinay 2007). That ecosystem services are too far removed from the language and understanding of bankers and analysts is a critical insight. It is ironic given that the use of the term ecosystem services was rationalized in part by ecologists and others because it was considered a more easily understandable abstraction (as I described in chapter 4). Indeed, if the turn to ecosystem services is for some too utilitarian and commodity focused, what this banker is suggesting is that the terminology is still not quite financial enough.

Challenge 2: Lack of defined unit of measurement or commodity

But for participants in these conferences the problem was not simply one of communication or language. As one participant stated in the London conference, biodiversity is "not by definition a sort of thing that is a commodity." She went on to say, "simplifying what is complex is really quite a challenge." But, she continued, "it is a challenge that we have to face." I described above conference participants' jealousy of the "CO_2 equivalent" unit that made carbon a "natural commodity" and made the carbon market operational. Another key challenge mentioned by speakers in London and New York – and a familiar refrain among proponents of biodiversity markets and valuation in general – is that biodiversity lacks a standardized unit of measurement, an equivalent to the "CO_2 equivalent." The IUCN representative emphasized this point in detail, comparing biodiversity to carbon and saying there was no agreement or coherence on how to measure biodiversity. In other words, there was a lack of standardization:

> There are so many values tied up in biodiversity – intrinsic and cultural values, the use values from ecosystem services to rural communities ... Now bringing together all those aspects of value – we are going to need to think about how to simplify that, how are we going to boil that down into indicators that can be used practically for management purposes.

Similar questions about what to measure in the production of credits exist in more targeted ecosystem markets, as Robertson's (2004, 2006) work on wetland banking demonstrates. Describing the problems of measurement, a New York participant explained how regulators in California were measuring each ecosystem function, including scarcity and rarity of various species. At one time, he said, "for the very rare or high-value types you could get twice as many credits, a higher number of credits than for more common ones." However, the Fish and Wildlife Service found the approach "too complicated" and so it went back to plain old acres.

A related problem concerns the spatial limitations of fungibility. While carbon markets and their participants and developers agree that a tonne of carbon here (for example, spewed by a car in Vancouver) is the same as a tonne of carbon there (for example, sequestered by the Amazonian rainforest), there is much concern about delimiting the limits of fungibility when it comes to ecosystems. There is no agreement, for example, that a hectare of temperate rainforest is the same as a hectare of tropical rainforest. Among proponents of offsets and mitigation banking, there is general agreement that these are usually going to be local, state, or, at most, national markets. As one participant stated, these limitations exist "for the simple reason that biodiversity is not fungible across ecosystems, across biomes. So typically, just as real estate markets are fundamentally local markets, biodiversity offsets are fundamentally local initiatives." The regional nature of biodiversity markets is often seen as a limitation in terms of scalability of the market and its growth, and also for the possibility of generating necessary international biodiversity capital. Therefore, while there are what we might call proper "biodiversity markets" emerging – say, for example, the growing species banks in the US and the offset markets elsewhere – a key issue is that they remain small and they are almost entirely domestic (or else voluntary, such as the Malua BioBank).[14]

A challenge with these small markets, however, is that the transaction costs are far too high to attract investment capital, a point reiterated over and over by financial actors at these conferences. As one speaker noted, "You spend as much money and time on a small project as you do a big project." The question that challenged many was how to aggregate several small projects and then securitize them to make funds worth the effort. For one banker, this may require "going across borders" of ecosystem types. Other participants saw room for financial product innovations for wetland and conservation banking akin to Real Estate Investment Trusts (REITs) that pool together properties and real estate assets. One participant stated, confidently, "We are going to see these developments" (meaning REIT-like structures). However, even those who had tried to bundle together a package of biodiversity products to dangle in front of investors – in one case, it was a combination of mitigation projects – noted that they were unsuccessful in finding any interest from institutional capital.

Challenge 3: Lack of "ecological-economic" data

A further challenge with the creation of biodiversity markets, as expressed by participants at these conferences, was a lack of scientific data. This lack begins at the most basic level; there is insufficient information about what biodiversity does and how it contributes to ecosystem services (see chapters 2–4). However, concern with this lack of information also illuminates an interesting difference between the New York and London conferences. In London, there was a general sense that, despite the groundwork laid by the Millennium Ecosystem Assessment, the basic scientific building blocks linking biodiversity and ecosystem services were inadequately developed. For example, a speaker from the Royal Society for the Protection of Birds highlighted recent research showing that places with high biodiversity do not always overlap with areas of high ecosystem service provision. At the New York event, however, there seemed to be less interest in the ecological basis of markets, and one participant even commented that he saw deals fall apart because ecologists were being too "detailed." While London participants seemed at least a little bit concerned with the dearth of accurate, applicable biodiversity data, those in New York seemed to prioritize simplicity and legibility for market production and expansion. My sense is that this difference might be partly explained by the fact that many at the New York event had experience with wetland and species banking projects (and the problems that stem from increased ecological complexity and detail in making them work) and by the fact that they had a close relationship to the approach as investors, business owners, or advocates.

The economic valuation data is even less forthcoming than the ecological. Several speakers in London made this point, noting that the Millennium Ecosystem Assessment barely scratched the surface of economic valuation. In part, this information gap was what led to the creation of the TEEB project, a global initiative, called for by Northern governments in 2007 to draw attention to the economic benefits of biodiversity. TEEB leaders were also present in London; they emphasized the need to produce some "sound numbers" that could quantify the contributions that biodiversity makes to economic growth and the costs from its loss (essentially reaffirming the aim of the TEEB project).

Expanding on this, as the IUCN representative noted, "We need to demonstrate the [economic] values [of biodiversity], use it to persuade policy makers to make changes, that in turn can mobilize real capital flows." But he explained the need for more detailed ecological-economic data to guide efficient investments, a mantra that reflects the outcomes of the Beijer Institute project on biodiversity (chapter 3). As he further explained, "It's obvious to anybody who works in markets and to most economists that without heterogeneous pricing, without differences in costs of compliance, you are not going to get trade." This speaker then displayed the famous mitigation cost curve from consulting giant McKinsey, which shows the low to high costs for greenhouse gas abatement from energy efficiency to forest conservation (see Figure 6.1). "We don't yet have a similar abatement curve for biodiversity and ecosystems," he went on to say. "But that is something we will need to develop." Such a curve would reveal the low-cost opportunities – that is, "the benefits the market is looking for at a competitive price." If biodiversity markets are unable to compare different courses of conservation action with others, they face significant challenges. As we have seen in previous chapters, calculating and modeling which parts of the ecosystem or biodiversity might be the "most efficient" conservation option is challenging and, in many cases, impossible. The ecological-economic information and measures needed to achieve such an abatement curve, and thus achieve efficient allocations of biodiversity, still do not exist.

Challenge 4: Lack of data, track record, and transparency

Quantification of the value of ecosystems and biodiversity is only one part of the knowledge gap for market-making. Some participants emphasized, further still, the lack of data on biodiversity markets – the lack of data about sales, revenues, costs, or even prices received for credits. As one participant in New York noted: "There is no track record. I have a track record of real estate sales in New York … books and books and books to

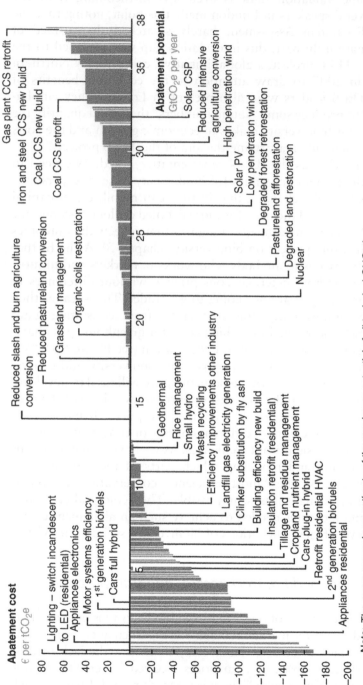

Figure 6.1 McKinsey greenhouse gas abatement curve, showing abatement cost beyond Business As Usual to 2030. © McKinsey & Company 2010 (by permission).

Note: The curve presents an estimate of the maximum potential of all technical GHG abatement measures below €80 per tCO$_2$e if each lever was pursued aggressively. It is not forecast of what role different abatement measures and technologies will play.

Source: Global GHG Abatement Cost Curve v2.1

understand how these things move." But, as he went on to say, "There is no such thing in the mitigation market ... No one has any idea of track record." So while the credit developer will tell the financier or investor that there is indeed demand for credits, there is little public evidence as to *when* the credits will sell. The same speaker stated, "Generally a credit that is on the shelf you can say with some level of confidence that it will sell, but you don't know when it will sell." The lack of information and "track records" means that most banks are not willing to lend to companies except against the raw land because, as one speaker in New York complained, "lenders do not understand the mitigation banking process." Difficulties obtaining financing are often compounded when, as sometimes happens, receiving regulator approval for a credit requires placing a conservation easement on the property and easements often devalue the property in the eyes of the bank.

According to conference participants, part of the problem stemmed from the lack of transparency in the market and the "tremendous reticence in industry to share the key data point, which is price" (i.e. of credits). Many at the New York conference saw increased transparency and information on price, volume, and turnover for credits as central to the growth in these new markets. As the boutique investment banker noted in New York: "In order for the asset class to get to institutional investments, you need to have a history, some data, a track record, statistics to show how performance [in that asset class] looks." Without this performance information, she stressed, you cannot compare between this asset class and others, which is what investors always do in a competitive market. Biodiversity remains an unknown. Without the evidence that it can earn high returns, or the evidence that it is low risk, biodiversity falls by default into the high-risk category. She went on to say that, for biodiversity to become mainstream, those involved needed to create evidence that it was a low-risk investment that also maintains solid returns: "It may be possible to get there, but it would require other stuff to help us do this: insurance, derivatives, contracts."

As I outlined at the start of this chapter, a senior advisor at a boutique UK-based investment firm began his presentation with questions about the very basic source of value in ecosystems and biodiversity. He began his presentation by asking: "What is being sold with a paper for ecosystem services. Is it land? A service? What is it that is being sold?" Across the Atlantic, the other boutique investment banker emphasized that even something relatively established, like wetland banking, faced these challenges:

> There is no bucket for the institutional sector to put this in. When you go to institutional investors they like to check their boxes where this fits. They say, "Where does this fit in my asset allocations scheme?" Is it

venture, is it alternative, is it real asset, is it something like this? This kind of looks and feels and quacks like real assets, but it is not really. So they say that it is a real estate derivative, and quite frankly, as of September, real estate derivatives are not the best place to be looking for money.

The challenge, as this participant summarized, is to make the "asset class" more fungible and common, to make "an asset class that institutional investors can look at and say 'this makes sense to me.'"

Challenge 5: Lack of regulation

For many at these conferences, creating robust biodiversity and ecosystem finance hinged upon developing the right governmental policies and institutions, the right regulatory environment. As one eco-entrepreneur explained in New York, market-makers "live and die by the regulatory environment." In the context of biodiversity markets, there is a large amount of political risk disliked by many investors. Referring to conservation credits, one wetland banking participant noted: "If the government says it has got currency, then it has got currency. They can change their mind – they are the government – and they have been known to do that. So, that is the political risk from the financiers' side." If no one is enforcing laws and policies related to mitigation and offsets, then, as one person noted, "there is no market."

Many participants at the conferences said that the inadequate regulatory environment was a key challenge for the emerging markets in biodiversity. In discussion, one person stated: "Basically you do not have an adequate policy environment that allows you to trade biodiversity and create value. We have to work on this. ... There is no market because there is no framework for such a thing." In other words, current regulatory frameworks for biodiversity markets were practically non-existent. Another participant noted the need for regulatory change: "If we want actual change you need regulatory change. That's what drives capital markets, that's what drives business." "Can we convince politicians to be brave," he continued, "and to open the door to new opportunities so that we can in fact generate the capital to conserve biodiversity?"

Discussions at these conferences starkly demonstrated the importance of governments in these markets. The general consensus among participants was that governments and regulators should be playing a critical role but were lagging, holding up the entire project of enterprising nature. There was a lone voice in disagreement to this point in New York: the investment banker. She wanted to be sure that the group understood that, while regulation can be good for the development of the market, "It does add to risk, and it adds to the time factor." She went on to say that this is "not necessarily a good thing" and that "sometimes a voluntary market that is driven by pure economics is a plus, because you don't have to deal with all sorts of other things." By "other sorts of things," she clearly meant the kind of regulatory frameworks and interventions that other participants were so clearly advocating. This participant held a more pure market view of how these biodiversity commodity-making practices should unfold. Hers, however, was a minority opinion in the larger discussions of the events, within which participants agreed that some government interventions were needed to establish frameworks for these new markets.

Developing agreement on governmental regulation and policy for market creation is even more challenging at the international level. In the international sphere, some pinned their hopes on the multilateral processes (e.g. United Nations Framework Convention on Climate Change (UNFCCC) and the Convention on Biological Diversity (CBD)), arguing that international government agreement was necessary for sizeable markets to emerge. Not everyone, however, saw the possibility for such an approach. At the more internationally minded London conference, one participant asked a panel of bureaucrats about the possibility for the CBD to "set up its own Kyoto-style framework to value ecosystem services in the way the UNFCCC has done for carbon." The room responded with twitters of laughter and the panels did not really even engage with the question, only noting that it would be up to the governments participating in those negotiations to advocate such a market framework. (See chapter 7 for analysis of the attempt to create, at the CBD, a mechanism for international tradable biodiversity credits.)

In spite of the lack of existing regulatory frameworks for biodiversity markets and the challenges of creating them at regional, national, and international levels, many participants articulated their belief in the need to move forward. Innovation based on partial knowledge and partial geopolitical agreement, they argued, had been a clear part of the creation of the carbon market – a history that they hoped to replicate in relation to biodiversity. The IUCN representative laid it on the table in London:

> We don't need to wait for everyone to agree. It would be lovely if the UN CBD was to agree to an adequate, credible financing mechanism [to]... meet the 45 billion dollars per year needed to deal with the protected areas gap. So we can move forward with those willing to take the initiative. We've seen with carbon markets that both on the regulatory side and voluntary side we can make progress and kick-start markets without getting bogged down by the process.

The idea that participants could avoid "process" in order to somehow "kick-start markets" is emblematic of the paradoxical optimism of the Green Power conferences, wherein hopeful, promissory language reigned, even within discussions regarding pervasive and fundamental challenges of implementation, particularly related to the crucial role of the government.

Biodiversity Markets: The Jalopy of the Financial World?

The challenges of biodiversity market making were revealed to me in my attendance at the Green Power events in London and New York in 2008–09. By sitting among the elites at the forefront of these market-making initiatives, I was able to gain a sense of the discourse that was developing among market makers and promoters of many stripes (or "species"). The Green Power conferences helped me to identify the array of actors involved, their ideas, and the "new domains of political-economic calculation" (Goldman 2005, 184) being formed and tested out in the realm of international biodiversity market making.

What I learned about this realm was that – in spite of the optimism of workshop participants, their hopefulness about the promise of this new form of bioeconomy – translating complex ecosystems and biodiverse spaces into units of tradable value turns out to be extraordinarily difficult. As I sought to identify and characterize the different "species" of actors involved in these market-making initiatives and to understand (and, later, track) the projects they proposed, I began to see that many projects stalled or were half-realized. Often these projects were success-ful in building new institutional infrastructure and creating new partner-ships between, say, a large international financial organization and a conservation NGO. Where they often ran into difficulty, though, was at the point at which biodiversity conservation might be made truly eco-nomically valuable. The challenges were legion, as those in the process of biodiversity market-making found that they lacked many things: shared language, a defined unit of measurement, scientific data, financial and regulatory understanding. At the "Biodiversity and Ecosystem Finance" conferences, participants described these challenges plainly, one practitioner, one investor, one manager to another.

These conferences show that making biodiversity conservation a site of accumulation is precarious and halting. It relies on large amounts of inputs and the development of extensive networks. While the idea that we ought to "sell nature to save it" is not new (McAfee 1999), and may be embedded within the concept of biodiversity itself (see chapter 2), what these significant challenges demonstrate is that the "neoliberalization of nature is far from complete, not without obstacles, and is anything but a smooth process" (Smith 2007, 21). The desperately sought patient capital continues to be elusive (see chapter 8).

Meanwhile, too, several large questions loomed over the entire project of hoped-for biodiversity market-making. Was the carbon market – the supposed model for a cap-and-trade program for biodiversity – functioning transparently, equitably, and effectively? Were existing biodiversity markets doing anything at all to abate biodiversity loss? Was any money actually flowing from these new biodiversity products? At the Green Power conferences, participants largely bracketed these basic concerns. All of this bracketing suggests that the act of market-making and the creation of biodiversity-related value – as an attempt to mainstream biodiversity conservation (see chapter 1) – was, through the diverse investments of a growing network of people, perhaps becoming an end in and of itself.

Finally, this chapter shows that several examples of biodiversity markets-in-making were also green development projects, initiatives that sought to create economic opportunities for people who live in or near the biodiverse regions that are meant to become a source of wealth. For example, in Tanzania and other developing countries, local people were paid for their work in the protection of watershed services. A Bio Assets initiative at the mouth of the Amazon River sought to create income-generating projects for local community members. In this way, the idea of creating value from biodiverse places was and continues to be linked to the idea of creating income not only for Northern investors but also for people in the Global South. These projects are the living, breathing face of what Kathleen McAfee (1999) terms "green developmentalism" or perhaps of "green improvement," to use the terms of Tania Li (2007). On offer in these projects is the promise of mutually beneficial improvement – North and South – achieved through neutral ecological-economic calculation and "fair" exchange: an axe, perhaps, for a bag of cash (recalling Figure 1.1 in chapter 1). In chapter 7, I continue this interest in "green developmentalism" and delve into an initiative that aims to make an internationally agreed upon biodiversity market: the Green Development Mechanism (GDM). This initiative aims to transform (very alive) "dead capital" into "live capital" in order to generate private investment in biodiversity conservation.

Notes

1   EKO is focused on bringing together "smart capital with people, projects, and companies that are poised to profit from new and emerging environmental markets (markets for carbon, water, and biodiversity)." EKO's investors were an elite group, including James Wolfensohn, former World Bank president and

Lord Jacob Rothschild (Sullivan 2010b). In 2015, EKO merged with Wolfensohn Fund Management to form Encourage Capital. Bayon is also editor and co-author of numerous books on mitigation banking, including *Conservation and Biodiversity Banking: A Guide to Setting Up and Running Biodiversity Credit Trading Systems* (Carroll et al. 2008).

2   In this chapter I draw from observations from these events, where I took extensive notes and recordings. Although these were public events and I enrolled in the event as a researcher, I do not attribute direct quotes by name, although where possible I link speakers to their organization or sector. In the chapter I also draw from the PowerPoint presentations, which were provided electronically to participants.

3   A growing literature exists on the creation of biodiversity markets and their political and ecological implications, beginning with Kathleen McAfee's (1999) article "Selling Nature to Save It." Critical scholars suggest that biodiversity markets are transformative of human–nature relations in a widespread and significant way. Arsel and Büscher (2012) see biodiversity markets as part of "Nature Inc.," a project in a "much larger, albeit uneven political economic project, that of establishing the supremacy of the logic of capital accumulation over society's relationship with nature" (58). See also Sullivan (2010b, 2013), Büscher (2009), the edited collection, *Nature Inc.* (Büscher et al. 2014), Büscher and Fletcher (2015), the articles in a 2012 special issue of *Development and Change* (Arsel and Büscher 2012, Bracking 2012), Brockington and Duffy (2010), Igoe, Neves, and Brockington (2010), MacDonald (2010), Robertson (2012), and Smith (2007).

4   See Sullivan (2010a, 118) for a discussion of how scarcity drives environmental market-making: "It is because of current environmental crisis that environmental health is now deemed valuable enough to invest in its commodification."

5   A policy-driven biodiversity market is a system created by governments (or other institutions) that creates payments or other incentives to landowners who undertake a particular type of land or ecosystem management. An information-driven market is a market in organics and other certified products (e.g. shade-grown agricultural products, bird-friendly coffee, ecotourism), where consumers create demand for "green products" and the private sector creates supply.

6   Wetland banking is an example of a government-created biodiversity market (see Robertson 2004 and 2006).

7   An update to the Ecosystem Marketplace *State of Biodiversity Markets* report found 45 existing mitigation programs worldwide, with another 27 in various stages of development (Madsen et al. 2011).

8   In spite of a lack of conclusive study of the wetland banking industry as a whole, several studies have suggested that wetland banking might not be resulting in positive outcomes for conservation. For example, analysis of one banking system, in Ohio, found that 12 of the state's 25 wetland mitigation banks were not up to scientific standard (Mack and Micacchion 2006). Scholars also raise questions about the cost-efficiency gains from market approaches that involve ecosystems where the commodity is location specific and non-fungible (Kroeger and Casey 2007, 324; Muradian et al. 2010, 1202).

9 The Equator Principles are a credit risk management framework for determining, assessing, and managing environmental and social risk in project finance transactions (Equator Principles 2013). Seventy-eight financial institutions are signatories to the Equator Principles.

10 The history of Bio Assets in this region is long and complex (Bio Assets 2015). A Japanese forestry company, Toyomenka ("Tomen"), acquired the land in 1969 and – after building a massive sawmill – sold the land to Bio Assets in 2001. Bio Assets partnered with a Dutch NGO to conduct diagnostic social and ecological studies (a "Social Economic, and Environmental Diagnostic" and a land use, land cover study). In 2003, Bio Assets signed an agreement with Petrobras to develop a pilot reforestation project; they began to develop understanding of the biomass and carbon data for 11 native species. Bio Assets further collaborates with Petrobras on community development initiatives in the area in a project called "Project Zero Hunger," which includes a 6 MW biomass plant.

11 REDD is a UN initiative that aims to create a fiscal mechanism that would pay countries in the Global South for carbon emissions reductions that result from avoiding deforestation and forest degradation (UN-REDD 2013).

12 On the community side, in 2008, Forest Peoples Programme, a UK-based NGO, examined the deal and found that Canopy Capital had conducted only weak consultation with the local communities in the forest (Griffiths 2008).

13 This speaker's analysis parallels the point made by Morgan Robertson (2006) about how wetland banking requires translation between relatively autonomous bodies of knowledge, each with its own logics and norms of verification.

14 The Malua BioBank is a forest rehabilitation project on Borneo that sells "Biodiversity Conservation Certificates" (BCCs); each BCC "represents 100 square meters of rehabilitation and protection of the Malua Forest Reserve" (Malua Biobank 2015).

7

Multilateralism vs. Biodiversity Market-Making

Battlegrounds to Unleash Capital

Rushing through the corridors of the airy Nagoya conference center at the 10th Conference of the Parties to the Convention on Biological Diversity (COP 10), on my way to yet another negotiation on the issue of financial resources, I came across a delegate unpacking a fresh box of very "hot pink" books. In bold color to distinguish itself from the mess of literature that adorns every flat space at every negotiation I've ever been to, the book, titled *The Little Biodiversity Finance Book*, grappled with the problem raised at every negotiation I've ever attended: that of cash flow.

The book suggests that over $300 billion is needed annually to conserve the world's biodiversity, to "maintain diversity in the human dominated context"; this figure rises to over $355 billion per year when climate change is factored in (Global Canopy Programme 2010a). Current funding for biodiversity conservation, the authors lament, is only about 10% of this – an estimated $36–38 billion per annum. In addition, half of these funds are delivered in the EU, US, and China, leaving large parts of the world with low or limited resources for conservation. The overall need, then, is for at least a ten-fold increase in conservation funding – or so say global biodiversity experts.

Environmental finance – the generation and distribution of financial resources for biodiversity conservation – is a central aspect of contemporary international environmental politics. The question of who will pay for conservation is perhaps the main sticking point in negotiations of the multilateral Convention on Biological Diversity (CBD); government

Enterprising Nature: Economics, Markets, and Finance in Global Biodiversity Politics,
First Edition. Jessica Dempsey.
© 2016 John Wiley & Sons, Ltd. Published 2016 by John Wiley & Sons, Ltd.

delegates consistently debate the quantity of funds for conservation and how those funds should flow and be governed.

The original articles of the CBD clearly articulate the differentiated responsibilities for meeting CBD objectives.[1] This was the promise of the 1992 Rio Earth Summit: technical and financial contributions from the North would enable sustainable, green economic growth in the South. In spite of the creation of a funding mechanism, the Global Environment Facility (GEF), intended to help developing countries cover the costs of biodiversity conservation, the difficulty of generating financial resources has plagued the CBD for years. The Global South repeatedly demands that the North increase its conservation funding, making good on this 1992 promise.

The North, however, continues to resist binding targets for increased public funds, and in the mid-2000s representatives of some Northern nations began to explore the options for "innovative financial mechanisms" (IFMs) to address this problem of lacked resources. This too was the subject of *The Little Biodiversity Finance Book,* and along with the book a new acronym was announced – "Proactive Investment in Natural Capital" (yes, PINC) – referring to the need for "innovative" options to finance conservation, such as payments for ecosystem services and environmental fiscal reform (e.g. subsidy reform), and biodiversity and carbon offsets. The press release that went along with the book suggested that government leaders assembled in Nagoya for COP 10 could raise $141 billion annually for biodiversity.

But in order to release this latent capital, many conditions must be created and maintained. At the most basic, one must produce something like biodiversity or even land as existing outside the economy, as say, an externality (see chapter 3). The very possibility of "getting something for nothing," or of creating money out of what previously had no value in exchange, as Timothy Mitchell (2007) explains, "rests on the notion that the market has an 'outside'" (261). The idea that we can create new sources of value or a new flow of capital "would make no sense without arrangements whereby things can be said to exist outside the economy" (261) – dead or defective assets that might exist as material wealth, but do not exist as capital. In making these arguments, Mitchell draws from his research into state-sponsored land titling programs promoted by the likes of Hernando de Soto, exploring the movement of an entity – in this case, land – from "outside" the market to "inside." He challenges the highly cited thesis of de Soto, who argues that the awarding of formal title enables landowners to use their property as collateral to borrow resources; the property thus becomes a "live" asset.

Attempts to make biodiversity a "living" asset are not new, but rather are present within the very origins of the concept of biodiversity (see chapter 2). And in chapter 6, we witnessed market promoters discussing and debating strategies for the creation and expansion of biodiversity and ecosystem financial opportunities. Attempts to make these commodities, I argued, can be characterized by what they lack: a defined unit of measure and the scientific and financial knowledge that might make this unit possible, but also a lack of institutional arrangements needed to grease the wheels of such markets and capital flows.

This chapter is all about an attempt to create an international legal framework to unleash "biodiversity capital." In the parlance of the CBD, this involves debates over innovative financial mechanisms (IFMs), debates that play out mostly over two international negotiations in 2009 and 2010: the Third Meeting of the Working Group on the Review of Implementation of the Convention (WGRI 3) in Nairobi, Kenya, and the 10th Conference of the Parties (COP 10) in Nagoya Japan. I attended these meetings, participating in both the official discussions and the activist activities organized around their edges, and conducted interviews with participants. Inside this chapter I sketch out the largely North-South struggles over innovative finance in general, and then focus on a proposed international market-based conservation instrument: the Green Development Mechanism (GDM). The GDM – advocated for by several European countries and NGOs at negotiations of the CBD – aimed to create an internationally approved biodiversity conservation "credit" that could be sold, thereby attracting increased private sector investment in conservation. For proponents, this new mechanism would help create the legal and regulatory conditions through which latent capital in the Global South's biodiversity could be realized – to smooth the path for private capital flows in conservation.

I borrow Mitchell's framing and argue that the purpose of the GDM is to move the asset of biological diversity – currently undervalued and "dead" in capitalist terms – "across a line from outside to inside the market" (Mitchell 2007, 248). In the case of IFMs for biodiversity, the "dead asset" – biodiversity – is very much alive, composed as it is of living beings. At the same time, the process of invigorating or capitalizing the "dead" in the GDM is similar to that of land titling or property-making. The purpose of the mechanism is to transform the value of biodiversity into an abstract form, turning its material wealth into abstract capital that can circulate. Like de Soto's solution to global poverty, which relies on land titling and private property, the GDM is an attempt to create an "apparatus of representation" that will, as Mitchell has it, "transform dead capital into live assets" (2007, 249). (By "apparatus of representation" Mitchell means all the components necessary to create private property and land titles, such as legal systems, land titling offices, and land surveys.)

The redrawing of this line between "inside" and "outside" the market, as Mitchell powerfully argues, is not a technical process that can simply create capital and unleash latent value. Rather, the placement of the line is better understood as a "frontier region," a kind of battleground, a site at which "new moral claims, ideas about justice, and forms of entitlement are forged" (2007, 247). In this chapter, I suggest that we view debates over innovative finance for biodiversity conservation similarly, as a site of political struggle, a struggle that I witnessed firsthand in my research. At the CBD negotiations, debates over whether biodiversity can be made into an internationally tradable commodity proved highly contentious: Southern signatories repeatedly pointed toward historical inequalities in global patterns of biodiversity use, challenged Northern signatories to live up to their commitments to pay for conservation, and raised concerns over the commodification of nature. Although the very idea of a "mechanism" suggests a technical solution, a way to bypass the fraught international politics that have long plagued the convention, the dream of a technical solution turns out to be highly utopian. Proponents of new mechanisms for international environmental finance encountered nothing but politics.

A warning: international environmental law and policy involves endless acronyms, meetings, and a slew of actors. And it can make your eyes glaze over. Part of my goal in this chapter is to familiarize readers with how these processes work, to show in some detail what geopolitical and epistemological power struggles look like in an arena taking place at such a distance from most of us – geographically but also via its technical language and obscure processes.

Battleground 1: WGRI 3 in Nairobi, Kenya

By the early 2000s, many organizations and governments were experimenting with international environmental market-making and new ways to generate much-needed conservation funds.[2] Within negotiations of the CBD, too, Parties began to explore new methods of resource mobilization – alternatives to the bilateral or multilateral funding programs derived from Article 20. We can track the Parties' movement toward IFMs through the Conferences of the Parties (COP) negotiations that take place every two years:

- In 2002, at COP 6 in Den Haag, Holland, Parties began to talk about the role of the private and financial sectors.
- In 2006, at COP 8 in Curitiba, Brazil, the Parties decided to conduct a review of the available financial resources for implementation of

the Convention. They also asked the Secretariat – the CBD's bureaucracy, headquartered in Montreal – to prepare a strategy for resource mobilization that would be considered at COP 9.[3]

- At COP 9 in Bonn, Germany, Parties adopted a strategy on resource mobilization; the decision noted the importance of innovative financial mechanisms, but did not specify the kind of mechanisms that might be pursued.[4]

In the decision at COP 9, the Parties asked the Secretariat to prepare another document, this time on "policy options concerning innovative financial mechanisms." The document was to include "inputs from regional centers of excellence in a geographically balanced way" and be forwarded to the Third Meeting of the Working Group on the Review of Implementation (WGRI 3) – the Nairobi meeting in 2010 that I describe in detail below. The Secretariat then produced a document and draft recommendations and also began collecting opinions on IFMs.

Also after COP 9, the government of Germany, in collaboration with The Economics of Ecosystems and Biodiversity (TEEB) initiative, financially supported an expert group meeting on IFMs that took place in Bonn in early 2010, an "International Workshop on Innovative Financial Mechanisms." Such expert meetings serve as mechanisms where international experts can vet policy in the intercession (the period between COPs). What is crucial to draw out about this meeting, but also about the operation of power in international law, is that these meetings are often funded via voluntary contributions of donor governments. By picking and choosing what kind of policy elaborations they fund, Northern governments are able to further define the normative agenda of the CBD. Germany and the Netherlands are major supporters of innovative financial approaches, and the outcomes of this meeting in Bonn would feed into official debates over policy in Nairobi, where Northern nations would seek to promote IFMs as a viable supplement to conventional, publicly funded approaches.

WGRI 3: An introduction

In the intercessional period between COPs several smaller meetings take place at which participants draft the text of decisions to be made by the COP. While ultimate authority for the decisions about international law and policy resides within the Parties at the COP, much of the framing and language of a particular decision is developed prior to the COP meeting in these smaller groups and negotiations.[5] One type of

Figure 7.1   UNEP compound. Photo by author.

"mini-negotiation" is a "Working Group" meeting. These meetings are usually one to two weeks long and are called for by the COP as needed. Here I discuss one such meeting: the Third Meeting of the Working Group on the Review of Implementation of the Convention (WGRI 3) in Nairobi, Kenya, where participants extensively debated innovative financial mechanisms.

The UN compound in Nairobi is a lush space filled with gardens and ponds (Figure 7.1), located in the rich suburb of Gigiri. The US Embassy for East Africa is located across the street, and both it and the UN are heavily guarded and protected by high fences. In May 2010, more than 700 participants attended WGRI 3, arriving each morning in a fleet of taxis. Participants predominantly represented governments from the North, but also UN agencies, intergovernmental and non-governmental organizations, Indigenous and local community groups, public sector research, academia, and business (Figure 7.2 and Figure 7.3). As is the norm within UN multilateral environmental agreement negotiations, participation was unbalanced, with many (biodiversity-rich) countries in the South like Madagascar and India sending one or two delegates to handle the entire range of issues, while Northern countries sent several delegates. Belgium, for example, sent 14.[6]

Figure 7.2   Delegates entering the UNEP compound. Photo by author.

Tensions around the conservation cash flow problem were evident right from countries' opening statements. Southern representatives made demands for binding commitments of funds from the North; Northern countries resisted these fixed targets. Brazil's opening statement set the tone. It was delivered by a key player in the debate over financial resources, Maximiliano da Cunha Henriques Arienzo, who reported that according to 94% of Parties the "implementation deficit" in countries' National Biodiversity Action and Strategies Plans (NBSAPs) was due to "lack of resources." Uncharacteristic in diplomatic negotiations, he then issued a challenge to developed countries: "Who," he said slowly, "will make declarations of binding provisions?" Following Brazil's position, most countries from the South (such as the Africa Group and India) referred to the lack of resources and the failure of countries to make good on paying for the full incremental costs of conservation in the original CBD text. While governments from the North did not deny the need for more resources, they disagreed with the idea of any specific commitment of funds. For example, in their opening statement on the issue, both New Zealand and the EU questioned the utility of quantitative targets. India, meanwhile, proposed a doubling of international financial flows to developing countries for conservation by 2020.[7]

Figure 7.3 Plenary Hall in Nairobi. Photo by author.

Representatives from Northern countries did not advocate the idea of any binding commitment of public funds. Rather, they suggested alternative approaches to meeting the resource demand: they turned to innovative financial mechanisms. For example, on the second day of the negotiation, a delegate from Switzerland explained: "We recognize the shortage of finance. Therefore we need to develop innovative mechanisms on a sound basis. Innovative financial mechanisms should be part of a broader strategy from private and public [resources], [and] public private partnerships. This should not of course impede or reduce funds from the public sector." In other words, according to this delegate, innovative financial mechanisms could be part of meeting the funding needs of global conservation.

Should Parties encourage the development of "innovative" financial mechanisms for international biodiversity conservation? More specifically, should innovative financial mechanisms be considered within the ambit of the CBD? These were key questions raised at the Nairobi meeting.

North vs. South: Innovative financial mechanisms at WGRI 3

At the Nairobi negotiations, discussions about innovative financial mechanisms focused on one document: "Policy Options Concerning Innovative Financial Mechanisms." (This is the document prepared by the Secretariat, as requested by the Parties in the official decision of COP 9.) In intergovernmental speak, the document was known as UNEP/CBD/WG-RI/3/8 and this was pronounced (over and over) in the negotiation as "UNEP – slash – CBD – slash – WGRI – slash – three – slash – eight" (Convention on Biological Diversity 2010c; hereafter CBD 2010c).[8] The document outlines policy options for innovative financial mechanisms, including payments for ecosystem services, biodiversity offset mechanisms, environmental fiscal reform (e.g. taxes and subsidy reform), and markets for green products. The document also considers how biodiversity could be incorporated into other fiscal means and measures like international development and climate change financing. As the document makes clear, innovative financial mechanisms do not only mean market mechanisms; a wide range of ideas and approaches are suggested.

Negotiations focused almost entirely on the short list of recommendations at the end of the document; the WGRI's purpose was to develop agreement on these recommendations that, later that year, would be forwarded to the COP for a final decision.[9] The initial draft recommendations are reproduced in full below (Figure 7.4); they call for Parties to undertake "concrete activities" to develop IFMs and mobilize IFMs to a level of at least 10% of total funds (CBD 2010c, 17). The recommendations also identify organizations that support work on IFMs, including "the Organisation for Economic Cooperation and Development, the Global Mechanism of the United Nations Convention to Combat Desertification, the Business and Biodiversity Offsets Programme, and the GDM 2010 Initiative"; this naming of institutions provides legitimacy that is critical for project fundraising and for working with national governments.

Debating innovative financial mechanisms at WGRI 3

The EU was the most supportive of IFMs, along with countries like Switzerland and Norway. New Zealand and Canada also expressed support, albeit less vocally than the Europeans. Supporters other than governments included organizations like OECD and NGOs like Forest Trends. The head of the Business and Biodiversity Offsets Programme – a multi-sectoral partnership between companies, financial institutions, governments, and civil society organizations focused on biodiversity offsets – was present for much of the negotiation. Most active was a three-person delegation from the organization Earthmind, present in Nairobi in order to advance the Green Development Mechanism (a specific IFM proposal that I discuss in depth in the following section).

In Nairobi, however, the tone of the negotiations on financial resources was decidedly undiplomatic, and antagonism increased over the week. Expressing concerns with innovative financial mechanisms were several delegations from the South, led by the Africa Group and the Philippines, as well as some NGOs like EcoNexus and Global Forest Coalition

The Conference of the Parties,

Recognizing the persistent shortage of financial resources available to support ecosystem services and underlying biodiversity, and that the achievement of the 2020 biodiversity target will be dependent upon the level of available funding at all levels;

Having benefited from the International Workshop on Innovative Financial Mechanisms organized in collaboration with UNEP-TEEB and with generous financial support from the German Government;

Noting the contributions in advancing innovative financial mechanisms from the Organisation for Economic Co-operation and Development, the Global Mechanism of the United Nations Convention to Combat Desertification, the Business and Biodiversity Offsets Programme, the GDM 2010 Initiative, and other organizations and processes;

Being aware of a wide range of available innovative financial mechanisms with promising potential of generating new and additional financial resources to the achievement of the Convention's three objectives;

Acknowledging that in addition to resourcing potentials, innovative financial mechanisms can be an important tool to transform modern economic systems in a way that will sustain ecosystem services and underlying biodiversity and promote green development;

Being determined to mobilize adequate financial resources at all levels as agreed to in the Strategy for Resource Mobilization adopted in decision IX/11;

1. *Agrees* that at least 10 per cent of total financial resources in support of ecosystem services and underlying biodiversity, with an initial target of annual additional resources of US$ 1 billion, by 2015, will be mobilized from new and innovative financial mechanisms at all levels;

2. *Adopts* the plan of priority action to promote innovative financial mechanisms as a contribution to the implementation of the Strategy for Resource Mobilization in support of the achievement of the Convention's three objectives;

3. *Invites* Parties and Governments and relevant organizations to undertake concrete activities for developing, promoting and adopting innovative financial mechanisms, as suggested in the plan of priority action to promote innovative financial mechanisms;

4. *Invites* competent international and regional organizations, *inter alia*, the Organisation for Economic Co-operation and Development, the Global Mechanism of the United Nations Convention to Combat Desertification, the Business and Biodiversity Offsets Programme, and the GDM 2010 Initiative, to support the implementation of the plan of priority action, and to collaborate with the Executive Secretary to organize regional and subregional capacity-building workshops on innovative financial mechanisms;

5. *Urges* developed country Parties to provide voluntary financial contributions to support the further work on innovative financial mechanisms;

6. *Decides* that each meeting of the Conference of the Parties will review and determine the level of funding that is needed from innovative financial mechanisms and which complements available resources of the financial mechanism;

7. *Establishes* an Executive Body on Innovative Financial Mechanisms to be in charge of mobilizing the determined level of financial resources from innovative financial mechanisms, including:

   (a) Make recommendations on the funding needs and requirements in relation to innovative financial mechanisms;

   (b) Advance global consideration of innovative financial mechanisms;

   (c) Promote allocation and distribution of financial resources generated from global consideration of innovative financial mechanisms, to national programmes on ecosystem services and underlying biodiversity, based on agreed transparent criteria;

   (d) Guide necessary implementation arrangements on promising innovative financial mechanisms;

8. *Instructs* the Executive Secretary to constitute a Financial and Economic Panel as a technical advisory body to support the Executive Body on Innovative Financial Mechanisms.

Figure 7.4 Draft recommendations for innovative financial mechanisms as first proposed in Nairobi at WGRI 3 (CBD 2010c).

Figure 7.5 Contact group negotiations in Nairobi. Photo by author.

(Lovera 2010). Many longstanding delegates found this unusual and a sign of a return to the more antagonistic politics of the past, with division largely organized along the North–South divide. Much of the debate occurred during a "contact group" (see Figure 7.5) which is a small negotiating group that proceeds without translation and therefore excludes many delegates (especially those from Latin America).[10] Overall, developing countries' representatives issued repeated calls for Northern nations to account for historical, ecological debts and to fulfill the Rio promise. As a delegate from the Philippines stated in Nairobi: "Developing countries should focus their money on fulfilling their contributions under Article 20. Why are we scrounging for other funds? It is because developed countries have not fulfilled their obligations." Delegates from Brazil, and others from the Africa Group made similar comments, viewing IFMs as an overt undermining of a key tenet of Rio, an admission of a lack of political will.

But Southern delegates had other problems with the proposed IFM text as well, including its roots in Northern expert circuits of power-knowledge and its possible negative effects, including the introduction of new market-related risks and potential impacts on Indigenous rights. Through these debates, we can see that any attempt to render biodiversity as a "commodity in waiting" or "just outside the market" is not simply a technical process or task. Rather, commodification of biodiversity raises complex, contested issues of development promises, neocolonial expertise, and concerns about potential negative ramifications from the policies themselves, concerns not only expressed by academics or radicals, but by government diplomats.

Other delegates from the South, particularly the Africa Group, raised issues with the reports on which the proposed IFM text was based. The initial draft decision recommended that governments use the outcomes of the expert group – the German-funded meeting in Bonn (described above). Southern delegates, however, raised concerns regarding the composition of the expert group, which was dominated by Northern participants and organizations. The EU, however, pushed for stronger language in the decision regarding the report, half-jokingly declaring: "We paid for it, as such we ought to use it." The Africa Group responded negatively and, reflecting the tone of the negotiation, actually asked if the EU delegate insisted on referencing it "because he [his group] paid for the workshop?" The EU delegate responded with anger and intimidation: "Just a minute. I am going to look you in the eyes ... When I said 'we,' I meant 'we' as a COP or as a Convention. Yes, Germany probably ... I did not personally pay for it."[11] What this demonstrated, beyond the parochial nature of the negotiations at times, was how the politics of knowledge and expertise were always at the fore of CBD debates, as delegates from the Global South were often incredibly aware of the "rule of experts" (Mitchell 2002), and the problems with neo colonial development policies and practices defined largely in the North.

Various Southern Parties raised questions about the potential negative aspects of innovative financial mechanisms, particularly market mechanisms. Argentina, for example, raised the specter of the recent financial crisis, referring to well-known problems with unregulated markets and new financial instruments like derivatives that "required ... enormous state intervention" to fix. The delegate noted that there are risks in markets "that might cause adverse effects," saying that "we need to undertake examination of regulation by the state of new markets." Others, like the main network of Indigenous representatives around the CBD, organized as the International Indigenous Forum on Biodiversity (IIFB), along with the Philippines, called attention to the need to ensure safeguards in market mechanisms, particularly for Indigenous rights. They suggested that the decision should have a paragraph calling on Parties and other organizations to "ensure that appropriate safeguards for the rights of Indigenous Peoples and local communities are incorporated." This language, however, was blocked by Canada and New Zealand, who did not accept language that refers to the rights of Indigenous Peoples in relation to the CBD.

Recommendations from the negotiation

Despite the conflict in Nairobi, the final recommendations were relatively in favor of new IFMs. As one analyst for the meeting wrote:

> Given the novelty of these proposed policy options and the cautious approach of most developing countries, WGRI 3 succeeded in overcoming resistance to starting a "global discussion" on such options that will continue to occupy the CBD's future agenda. With delegates agreeing that innovative mechanisms, if established, shall be supplementary to the financial mechanism of the Convention, the comfort in exploring options may actually have increased. (IISD 2010)

Parties reached general consensus on the place of innovative financial mechanisms with an unbracketed decision that "encourages" Parties and other relevant organizations "to undertake concrete activities for developing, promoting and adopting innovative financial mechanisms, including the examination of the report of the International Workshop on Innovative Financial Mechanisms."[12] The decision also noted that these new and innovative mechanisms were supplementary to public, North-to-South transfers, appeasing those worried about IFMs supplanting public commitments. Brackets, the sign of disagreement, occurred in only a couple of places in the text, including around the paragraph that called on Parties to ensure safeguards for the rights of Indigenous People and local communities. Also bracketed was reference to two specific initiatives, the Green Development Mechanism and the Business and Biodiversity Offsets Programme. This bracketing remained due to concerns from the Africa Group over the initiatives' mandates, governance structures, sources of funding, funding criteria, and beneficiaries. These delegates were very aware that the conditions within which the knowledge was produced could shape the text and the final decision, and were – through their resistance – pushing for more inclusive approaches to generating solutions to problems faced by the CBD.

Promissory Mechanisms: The Green Development Mechanism

The Green Development Mechanism – a specific proposed IFM – was a source of controversy at the Nairobi meeting. The Africa Group, for example, took particular issue with the mechanism (and with the Business and Biodiversity Offsets Programme, another proposed IFM); one delegate stated, "We don't know anything about these projects; they are new institutions." The Filipino delegate paralleled this sentiment: "As Asia

Pacific groups, we are concerned that initiatives like this are on the table and we don't understand them." Parties from the Global South raised concerns about endorsing initiatives that they had little knowledge of and that were largely generated and developed by Northern organizations. As a result of these concerns, at the end of the Nairobi meeting, the GDM text was bracketed. The initiative nevertheless emerged as a specific, tangible policy option and was given a more active role in the recommendations. In fact, despite being in brackets, the GDM text developed beyond a mere mention: delegates added a whole paragraph calling for the development of this mechanism, and it would go on to be debated in Nagoya.

From where did this proposal of a specific IFM arise and how did it gain support? A close study of the GDM reveals it as the brainchild of a Dutch bureaucrat, circulated and promoted extensively by well-connected experts between 2009 and 2010. Soon, the GDM idea gained institutional support from government, international institutions, and NGOs, developing a wide network of allies. The GDM is a bold and, I would argue, even utopian proposal; its advocates are proposing, essentially, a land use planning process for the entire globe. The mechanism is premised on a belief that the action of seeming to move resources from the outside to the inside of the market (Mitchell 2007) can create "money out of nothing." Between 2009 and 2010, the GDM begins to appear to its growing network of supporters as a technical solution to the ever-political problem of conservation finance.

What is the Green Development Mechanism?

The Green Development Mechanism aims to transform the "inert" value of biodiversity in the Global South into a live, capital-generating asset that can be sold, generating much needed resources for conservation. As Francis Vorhies, executive director of Earthmind and lead proponent of the GDM explained in a side event in Nairobi:

> If we are serious about this ten-fold increase [in conservation finance], we need to figure out how to get the market involved. And the whole proposition of the GDM is to say, can we bring [in a] mechanism that gets private sector players, producers and consumers, investors, companies and so on, actively engaged in putting their money where their mouth is, and getting involved in market processes for conservation?[13]

The general idea behind the GDM is to develop an intergovernmental-approved asset in "effectively managed space." In other words, spaces that are managed for biodiversity conservation can generate "conservation plans" that can be sold. Vorhies explained the nature of the proposed commodity:

> So what are we supplying? For carbon we have tonnes. And carbon is like frozen orange juice or rice or coffee, a commodity, and we sell it in Chicago as a commodity. We can't do that with biodiversity. We can't talk about tonnes of nature, because nature is diverse. We've got that the coral reefs are different than rainforest, are different than grasslands, different than mangroves. Again, very simplistically … the simple thing we deal with in biodiversity is space. We sell a certain part of space that is effectively managed. And what do we call that space – area. And how do we measure it? A hectare? An acre, a square meter, or square yard.

The GDM, then, aims to create a standardized commodity in "well-managed space," similar to how the grain traders in Chicago created a standardized qualities of grain (Cronon 1992) or how the carbon market makes a standardized, internationally recognized unit of carbon sequestration (Lohmann 2009; MacKenzie 2009).

Vorhies summarized the concept in an interview with journalist Stephen Leahy: "The way it would work is that any entity – an organization, a local community, etc. – could create a ten-year plan for sustainable use in a specific environment: forest, wetland, coastal region, or coral reef … After an independent audit, the conservation plan would be certified and its 'biodiversity-protected hectares' would be put on the market for purchase" (quoted in Leahy 2010).

What is being sold, then, is not the object of biological diversity, but rather the right to a particular kind of management on the landscape. Credits would be purchased, according to GDM promoters, by private sector actors looking to fulfill their mandates for corporate social responsibility or even by members of the public wanting to buy a hectare of rainforest or coral reef that they know will be properly managed and conserved. As Vorhies explained: "If the top 500 companies globally were to commit just one hundredth of one percent of their annual revenues it would generate 2.5 billion dollars for conservation annually" (Leahy 2010). The GDM is a mechanism that can bring biodiversity "inside" the market, allowing, as Vorhies explained in Nairobi, "the market to get involved more substantively in implementing the CBD."

The roots of the GDM: The "biodiversity expert network"

The concept underlying the GDM was initially conceived in the Netherlands Ministry of Housing, Spatial Planning, and Environment. At COP 9, this ministry hosted an event to explore the concept that was to become the GDM, moderated by a Dutch bureaucrat, Arthur Eijs, a major supporter of the GDM.[14] The summary notes from this event state: "The most important conclusion drawn by moderator Arthur Eijs was the level of (critical) support for thinking along the lines of an international instrument. It was found that it could significantly add to existing funds for biodiversity conservation and truly tap into a new market by mobilizing private capital" (Balancing Biodiversity 2008, 2). The Dutch are recognized advocates in the area of biodiversity, finance, and the private sector; within the Netherlands, support for economic mechanisms for biological diversity goes to the highest political levels. An April 2009 memo from the Minister of Housing, Spatial Planning, and Environment and the Minister of Development Cooperation states that a long-term goal of the Netherlands is preventing "further loss of biodiversity at home and abroad."[15] The memo specifically mentions the GDM as one possibility for internalizing externalities caused by the "activities of the Netherlands."[16]

An interviewee said that while "the Dutch were the ones who had the idea and decided to set up an initiative," they "very quickly" began speaking to "their partners" to form a small committee. This committee included representatives from the global biodiversity expert network: staff from the Secretariat to the CBD, UNEP, the OECD, Joshua Bishop from the IUCN and, of course, Arthur Eijs from the Dutch government. According to my interviews, the International Finance Corporation also played an important role at the outset, as well as representatives from other organizations like the Business and Biodiversity Offsets Programme and Forest Trends. Many of these representatives have collaborated in the past; one interviewee referred to the group as "old friends." Over the next year and a half, a number of individuals, institutions, organizations, and academics collaborated to advance the GDM. A series of international workshops enabled actors to come together and build consensus.

In February 2009, only nine months after the previous COP, a small expert workshop was convened under the title: "A Green Development Mechanism: Towards an International Market-Based Instrument to Finance Biodiversity Conservation." Held in Amsterdam, the meeting was funded by the Dutch government and the International Finance Corporation, and its participants sought to explore the viability of the GDM in possible forms and generate plans of work for its further development (Mullan and Swanson 2009a, n.p.). The background material and final report for the workshop were created by two academics: Katrina Mullan, then a PhD student in the Department of Land Economy at the University of Cambridge, and Tim Swanson, an economics professor at University College London.[17]

A total of 38 people participated in the workshop: 30 from the North and 8 from the South.[18] For a meeting that sought to discuss an international mechanism for green development, the overwhelming number of participants from Europe and North America was surprising. The ratio represents a strategic error on the part of the organizers – at later negotiations, including in Nairobi, GDM proponents would find it difficult to get "buy-in" on policy proposals so clearly formulated by experts from the North.

Operationalizing an "ecological-economic tribunal" for life on earth

Proponents of the GDM are driven by two interlinked rationales. First, proponents sought to create a long-term incentive for the values of global biodiversity conservation that can outweigh the "local" incentives of land conversion for commodity production (e.g. soy, palm oil, etc.). They want to capture the global public good value that biodiversity has for ecosystem resilience, which is "fundamental to life" (Mullan and Swanson 2009b, 5). What's needed, they argue, is a permanent change in the system of incentives, wherein payments for the global benefits of biodiversity could be used to change "local decision-makers' land-use calculus" (Mullan and Swanson 2009a, n.p.). If one wants to make biodiversity conservation "competitive" with other land uses "like deforesting and planting palm oil," one representative from a major development bank said in an interview, "you have to artificially do that right now." To make her reference point totally clear, she went on to say: "in terms of protecting large chunks of nature, what needs to happen is an agreement à la Kyoto, or à la something."

The second rationale for the GDM was the internalization of economic externalities and, ultimately, full-cost accounting. The background technical GDM document specifically lays out this rationale, citing previous mega-reports from the Millennium Ecosystem Assessment (discussed in chapter 4) and The Economics of Ecosystems and Biodiversity (TEEB). Both of these source reports identify "a lack of markets and prices for ecosystem services to be among the underlying causes for continued biodiversity decline, and both call for further development and implementation of market based instruments to 'internalize' ecosystem values in economic decision making" (Mullan and Swanson 2009b, 2). This "internalization" is envisioned as the ultimate way to alter the incentive structures that cause land conversion.

Internalization was a critical part of the vision of the Dutch government, which wanted to ensure that decisions in the Netherlands incorporated the full cost and externalities of their operations, as related to biological diversity. One proponent of the GDM told me that "governments want to undertake activities to pay for the full costs of their production and consumption, but there is no mechanism by which they can do that." Animating the GDM is a vision of a more perfect market where all actors pay the correct price for biodiversity and the services biodiversity provides. That the Dutch would like to rein in their impact on the planet, impacts far removed from Dutch people, is a far-reaching and even radical notion.

The dual rationale for the GDM is exemplary of what McAfee (1999) famously called "green developmentalism," focusing on the promise of genetic resources in the early days of the CBD (see chapters 2 and 3). Such an approach to development focuses on generating market solutions to environment-development problems, "promoting commodification as the key both to conservation and to the 'equitable' sharing of the benefits of nature" and "enlisting environmentalism in the service of the worldwide expansion of capitalism" (McAfee 1999, 134). The promise of the GDM is that – once the right conditions are created – biodiversity-rich countries will be able to unleash their superpower status through conservation. (One United Nations Development Programme (UNDP) initiative claims, for example, that Latin America is a "biodiversity superpower endowed with the greatest amount of natural capital in the world" (UNDP 2010).) In this reimagining of the world order, the globe's massive inequities are corrected by not only bringing traditional commodities like soy, timber, or gold into market exchange, but also by making the conservation of biodiverse spaces profitable for countries in the Global South. If these countries can bring the "latent values" in biodiversity – its genetic resources, the ecosystem services it supports – into the market, the reasoning goes, then they can achieve green development. The promise, then, of the GDM lies not only in generating funds for conservation, but in reconfiguring markets in a way that will address historic imbalances in the global allocation of benefits from nature. This utopian vision, however, encounters both practical and political challenges.

The function of the mechanism and the limits of fungibility

"The GDM would not be a fund," explained Vorhies at a Nairobi side event. "It would be a market certification/verification mechanism" like the Clean Development Mechanism (CDM).[19] The background paper for the mechanism identifies the four key aspects of the proposed approach: an aggregate constraint on development activities, a certifying authority and certification process, an agreed-upon standard or procedure for trading development rights, and an effective system for monitoring and enforcing the mechanism (Mullan and Swanson 2009b, 3). The aggregate constraint (or "cap") on the conversion of natural systems would create a "demand for rights to develop" – a demand that could be met by buying "conservation credits." It is this demand for development rights and willingness to pay for them that create the mechanism for the fiscal transfer of funds from the North – those obligated to buy development credits – to the Global South – those doing the conserving.

Definitions of the "constraint" and the methodologies and criteria for a "certified credit" are linked to prioritizing space; they are deciding which natural systems should remain unconverted and which can be "spared." As the report from the Amsterdam workshop states:

> All of the proposed GDM designs involve implicit or explicit decisions about what should be protected. Therefore it is important to consider how this prioritization should be carried out (e.g. it could be done through the use of spatial mapping and planning). The nature of the decision-making process should also be considered; i.e. whether it should be based primarily on scientific assessments or on broader political processes. (Mullan and Swanson 2009a, n.p.)

The vision of the GDM presented within these background reports is so removed from political reality it makes my head spin in disbelief. As I have mentioned, were the GDM to be fully realized, it would result in a comprehensive ecological land use planning process for the entire globe (see Mullan, Konotolen, and Swanson 2009). This kind of panoptic vision, the endless pursuit of a permanent system of knowledge in which all human–nature relations might be managed in a "rational" way, is reminiscent of the work of the Beijer project (described in chapter 3), and Hayekian dreams of "catallaxy": a peaceful world order that would result from a perfectly operating market based upon perfect knowledge. Key social and political challenges are immediately apparent. How would GDM proponents get intergovernmental agreement for such a plan? How would local communities and Indigenous Peoples who use certain land areas be compensated? Even in advance of these questions of legal process and rights, however, GDM proponents encountered the question of fungibility.

As I described in chapter 6, biodiversity cannot be made "universally fungible" (the way carbon is under the CDM). The authors of the background technical report on the GDM write: "while benefits of emissions reductions are the same regardless of the source, the value of biodiversity conservation is highly specific to its location and the precise type and condition of habitat conserved" (Mullan and Swanson 2009b, 10). Biodiversity's site specificity means that any mechanism that offsets faces the challenge of verifying the "losses of biodiversity in one location" with "increases in another location" (Mullan and Swanson 2009b, 10). How can one "fully account for" the losses of biodiversity in one site and make them equivalent to the "gains" in biodiversity conservation in another?

As one attendee asked in a GDM presentation in Nairobi: "Why is this considered a viable option if biodiversity cannot be traded viably on a global scale, especially when ... 'areas of biodiversity importance' are not tradable across different regions of the world?" She went on: "There is room for additional cash. The risk is when you say, well, you can destroy over here in the US, and then buy some coral reef in Indonesia." One academic I interviewed within the course of my research suggested that establishing biodiversity-oriented trading schemes means, in terms of fungibility, that "you've got to be willing to accept some pretty ugly trade-offs." "These trading markets are not trading apples for apples," he said. "They're trading proxies and a lot of the programs aren't being honest about that."

Back in Nairobi, Vorhies answered that the GDM is not "an international offset mechanism." However, Vorhies claimed, while the GDM is not a "like for like" mechanism, biodiversity credits might be purchased to deal with the impacts in the "broader value chain":

> Offsets are something that must be done locally ... if a company takes five hectares and turns it into a factory site, then they need to offset it with the same type of ecosystem. They cannot do [it with] five hectares across the planet ... But this factory has stuff coming into it, and coming out. So there is a value chain that the localized biodiversity offset does not capture ... So the GDM could say, in addition to the localized offset, which should be done, maybe a company could support conservation to deal with the broader value chain.

In other words, while recognizing the impossibility of biodiversity offsets outside of ecosystem type, Vorhies believes in the possibility of fungibility (to some degree) in the process of internalization for the "broader value chain." For example, if the factory produces candles that use palm oil, it has impacts on biological diversity because of palm production (which is a leading cause of deforestation), impacts that could be internalized via a purchase of a credit through the GDM. In this way, a GDM can be part of internalizing externalities across sites and spaces and ecosystems presumed to be "non-fungible." In other words, according to GDM proponents, the idea that we cannot swap "biodiversity loss/degradation here" for "biodiversity protection there" is a limitation not based in the actual "materiality" of biodiversity – the fact that it is not the same here and there – but rather on a social and political decision or agreement that these sites are "non-fungible." As a kind of work-around to this non-fungibility, the GDM aims to create exchange more akin to barter trade, or barter trade plus, where "equivalence" is established not on the basis of a universal abstraction (i.e. socially necessary labor time or, say, eco-system service quantity), but rather on the assessments of transacting parties. Equivalence under this mechanism would be socially and political negotiated and agreed upon between parties, including perhaps govern-ments. (That said, a principal difference from barter is the exchange of cash for the biodiversity conservation secured.)

Global promotion of the GDM

A key outcome of the February 2009 meeting was a generalized agreement to move ahead with some sort of GDM. Earthmind and other GDM proponents took the idea on the road, presenting it at workshops and events around the world in order to obtain political support before COP 10 (see Table 7.1). In a submission to the GDM steering committee, Francis Vorhies set out the challenge in a jumble of acronyms of interna-tional bodies and organizations:

> What are the points of engagement along the way with respect to not only CBD events, but also relevant events of other biodiversity-related MEAs, as well as other processes/institutions such as UNFCCC, CSD, WTO, WB, OECD, TEEB, GEF, etc., etc.? To get something developed, on the agenda and backed by the Parties at COP 10, will require a lot of work before then.

The GDM initiative went into full gear with the goal of trying to have a GDM adopted by Parties at COP 10 as an innovative financial mecha-nism "in development."[20] As proponents presented the GDM at many events around the world, the idea appeared to be gaining traction with some members of the international conservation community, as well as the international business community. (For example, GDM proponents were successful in having the mechanism mentioned in a charter that came out of the Third Business and the 2010 Biodiversity Challenge Conference, held in Jakarta.)

What was presented at these events, however, was much more limited than the vision elaborated in the background documents of the first discussion of the GDM. The need for an aggregated constraint – a cap – largely disappeared. The focus became, instead, obtaining political buy-in for the general idea of a voluntary GDM and creating a standard

Table 7.1   Meetings at which the GDM initiative is present and active.

Month and Year	Event, Location
May 2008	COP 9 side event "Balancing Biodiversity," Bonn, Germany
February 2009	First meeting for GDM, Amsterdam, the Netherlands
July 2009	OECD meeting, Paris, France
August 2009	Public Seminar on Biodiversity, Economy and Business: "International Market-based Instrument to Finance Biodiversity Conservation," Tokyo, Japan
October 2009	Sustainable Agriculture Initiative (SAI) Platform, Rotterdam, the Netherlands
October 2009	UNEP FI Global Roundtable, Cape Town, South Africa
November 2009	TBLI CONFERENCE™, Amsterdam, the Netherlands
November 2009	Boosting Investments in Biodiversity and Ecosystem Services, Amsterdam, The Netherlands
November 2009	The Business of BioTrade: Conserving Biodiversity through Using Biological Resources Sustainably and Responsibly, Geneva, Switzerland
November 2009	Third Business and the 2010 Biodiversity Challenge Conference, Jakarta, Indonesia
January 2010	International Workshop on Innovative Financial Mechanisms, Bonn, Germany
February 2010	Second workshop on a Green Development Mechanism (GDM), Jakarta, Indonesia
March 2010	World Business Council on Sustainable Development (WBCSD) workshop on TEEB, GDM, and COP 10, Montreux, Switzerland
April 2010	Geneva Trade and Biodiversity Day, Switzerland
May 2010	WGRI 3, Nairobi, Kenya
June 2010	Seminar on GDM, Tokyo, Japan
September 2010	Meeting with representatives from the Africa Group, Geneva, Switzerland
September 2010	Biodiversity and the Fight against Poverty: Opportunities for Africa, Libreville, Gabon
October 2010	COP 10, Nagoya, Japan

approved "biodiversity credit." The goal became less about full-cost accounting and more about gathering political support (and funds) to further develop the GDM. In one presentation (to the UNEP Governing Council), a GDM proponent stated that the aim was "to mobilize new and additional resources in support of the implementation of the CBD, especially from the private sector." The demand for such biodiversity credits would not be created by establishing a cap, or constraint on development, but rather would begin with establishing "a crediting scheme to identify and verify the biodiversity and/or development outcomes of [a] project" (Green Development Mechanism 2010a; hereafter GDM 2010a). This would require a "GDM standard and related methodologies and modalities for auditing, verification, certification, monitoring and reporting" (GDM 2010a). In other words, the GDM would begin not with gaining full intergovernmental consensus on a cap, but rather with developing agreement on how credits would be defined; it would begin not with creating a constraint on land use change, but with the creation of the commodity. This is, essentially, creating a trade without creating a cap.

A conference room paper: How the GDM entered official COP text

At the negotiation in Nairobi, the draft decisions (originating from the CBD Secretariat) referenced the GDM initiative, but did not call for direct support or elaboration. Earthmind, however, arrived in Nairobi armed with specific language about how the Parties could, beyond mere reference, support the concept in order to "commence the initiation of work under the CBD toward establishing a GDM" (Green Development Mechanism 2010b; hereafter GDM 2010b). Earthmind circulated a document encouraging Parties to adopt a standalone, action-oriented paragraph on the GDM within the innovative financial mechanisms discussion. They suggested that the Parties should officially propose a large-scale, multi-institutional dialogue to set the GDM in motion, calling for

> Competent international and regional organizations – including the CBD Secretariat, Earthmind, IUCN, OECD, and UNEP – to establish a global discussion on the need for and modalities of a green development mechanism, which, in its pilot phase, would develop a voluntary standard and certification process for validating the supply of biodiversity-protected areas and a market-based institutional framework for enabling payments by companies, consumers and other stakeholders (GDM 2010b).

The organization also organized two side event presentations to introduce the concept, outlining both the concept and the way forward.

As I have already mentioned, some Parties in the negotiation, particularly the Philippines and the Africa Group, viewed the GDM with suspicion. One delegate from Malawi stated that the initiative was "not clearly known to Africa and their mandate is not clear." Simone Lovera, a representative from an NGO called the Global Forest Coalition, wrote an article in *ECO*, the civil society newsletter that strongly criticized the GDM. The GDM, wrote Lovera, is "a classical example of an idea that was developed by some young academics with little clue about the latest developments in international environmental policy-making" (2010, 3). She decries the lack of a cap or targets in the biodiversity regime, arguing that "trade without a cap would be a classical example of trading in hot air" (2010, 4). But for Lovera, the worst aspects of the GDM were that it "has been developed for a planet where biodiversity is carefully separated from people" (2010, 4). "The entire design document for the GDM," she continues, "does not include ANY reference to the rights and needs of Indigenous peoples, women, farmers, or local communities. Such exclusions leave these people and their lands open to land grabbing and dispossession" (2010, 4).

When such concerns were raised and circulated widely at the negotiation, the Earthmind lobbyists stepped into action quite literally, getting up from their chairs to walk over and talk with their allies in the government and try to convince the skeptics. Lobbying at international negotiations is incredibly dependent on physical presence, as people must talk to delegates and convince them of their trustworthiness, of the sound nature of their arguments and evidence. It is often easy to see peoples' perspectives on an issue, particularly when NGOs are involved; when critical comments are made, eyes dart around the room, and people get up and whisper in ears; they pass papers back and forth. As a part of the civil society collective producing the *ECO*, where Lovera's article appeared, I received a stern talking-to by several members of the GDM network, both governmental and non-governmental.

Despite these concerns, the first revised paper on IFMs gave the GDM a more active role in the text. A new paragraph appeared in what is called a "conference room paper," which is a revision of only the decision or recommendation part of the documents. According to the new, additional paragraph, the Conference of Parties "Invites interested organizations and initiatives to consider the need for and modalities of a green development mechanism, which, in its pilot phase, could develop a voluntary standard and certification process for validating the supply of biodiversity protected areas and a market-based institutional framework for enabling payments by companies, consumers and other stakeholders." The language was almost word-for-word the language prepared by Earthmind.

How did this happen? Inserting language in the text requires the support of a Party, a government signatory. So how did the words of one organization become so directly integrated into a draft international legal decision? The story is not straightforward. The co-chairs and the Secretariat are responsible for preparing the conference room papers (CRPs) on the fly, usually late at night. Since one staff member of the Secretariat actually sat on the steering committee for the GDM, it was clear that there was at least some support there. However, most critically, there were a few government delegates who strongly support the GDM, particularly the Netherlands, which financed the initiative, but also other countries like Switzerland and Germany.

In response to the revised paper with the expanded text on the GDM, the Africa Group continued to insist on the "bracketing" of the GDM and the Philippines asked for all the references to be deleted, continuing to state their unfamiliarity with the mechanism. Other countries, including Canada, also suggested deleting references but only in order to get rid of the brackets, as bracketed text creates more work at the next negotiation. The EU and Switzerland insisted on keeping references to the initiative in. The final result was that all references to the initiative were bracketed, including the full new paragraph, but nevertheless forwarded to COP 10, which took place less than five months later, in Nagoya, Japan. The GDM proposal, initiated by Dutch bureaucrats, promoted vigorously worldwide by Earthmind and other supporters, and vociferously challenged by many Southern representatives, was now officially on the table at a Conference of the Parties to the CBD.

Battleground 2: COP 10 in Nagoya, Japan

Five months after the Nairobi working group, the debate over IFMs was underway again, this time in Nagoya. Japan's third-largest city was home to the 10th Conference of the Parties, which took place in a breezy, open, and sprawling conference center (see Figure 7.6 and Figure 7.7, below). Over 18 000 people attended, including 122 ministers, 120 parliamentarians, 200 mayors, and 650 heads of municipalities. The event received a fair amount of media attention in Japan as well as abroad. On the issue of financial resources, the cast of characters shifted slightly, but importantly, from earlier negotiations. There were constants: the lead negotiator for the EU returned, amid rumors of censure for his aggressiveness in Nairobi, as did the negotiators from Brazil and Switzerland, who played active roles in the debate. The negotiation tone changed dramatically, however, largely due to the presence of two negotiators from the Global South, who appeared on the scene with strong mandates: Bernarditas de Castro-Müller (from the Philippines) and Carla Ledezma (from Bolivia).

Figure 7.6   Nagoya conference center. Photo by author.

Third World women interruptus, redux

Bernarditas de Castro-Müller arrived as a seasoned negotiator in the multilateral scene, a Filipino delegate considered "a genuine diplomatic celebrity," apparently known to her friends as "Ditas" and her enemies as "dragon woman" (Harris 2009). She became famous for her work in the climate finance negotiations, where her strong positions led to her dethronement as lead negotiator for the Philippines, supposedly at the request of the US (Harris 2009). Unlike in Nairobi, in Nagoya de Castro-Müller negotiated in the financial resources discussion often on behalf of the G77 plus China.[21] As she repeated over and over, the issue of financial resources is central to the entire Convention. "You need timely, adequate, and predictable funding and capacity building," she said. "This is a precondition for anything here." In the course of the

Figure 7.7   Inside the ministerial plenary at COP 10. Photo by author.

negotiation, de Castro-Müller continually raised the issue of debt load for Southern countries.[22] She also brought a wealth of experience with the CDM, and this experience made her very suspicious of the GDM. In the negotiation, referring to the new paragraph calling for the development of the GDM, Müller stated that the problem with this paragraph "is that they already show the way to a specific financial mechanism." She opposed this, because "it looks, talks, walks like a duck, and that duck is the CDM." The failures of the CDM (as described in chapter 6) were – for de Castro-Müller – enough of an argument. She ended her intervention with a note of weary finality: "I will not go into this."

Carla Ledezma, from Bolivia, was the second new influence in GDM negotiations. She represented the vastly expanded presence of the ALBA grouping (a collection of socialist and social democratic governments from the Americas) and played the lead role in the ALBA's financial resources negotiation.[23] Bolivia took the lead in drawing attention to the negative aspects of market mechanisms. The country was newly empowered after the successful "World People's Conference on Climate Change and the Rights of Mother Earth" in April 2010. The conference was attended by over 30 000 people from governments, NGOs, and social movements, and focused on addressing the "root causes" of climate change. The People's Agreement emerging from the event rejected what the authors call "false solutions" to climate change, including large-scale biofuels and geo-engineering, but also market mechanisms such as carbon offsets.[24]

Opening statements

In Nagoya, the opening statements on financial resources retrenched the divisive North–South positions formed in Nairobi five months earlier. Kenya called for a commitment "to increase funding ten-fold, with sharp and focused targets," and South Africa reiterated the need for any innovative mechanisms to be supplementary, urging "the North to provide in accordance with historical obligations." Meanwhile, Northern countries reminded delegates about limitations of public finance. Norway referred to the green economy and the need to experiment with the "most innovative" mechanisms, mentioning payments for ecosystem services and offsets that could attract new sources of funding. The delegate stated, "We agree that a substantial increase in funding is needed, however, we have to take into account that resources are not easy in the future."

Bolivia, however, raised the stakes, claiming that financial resources should come from public funds and rejecting market mechanisms. The delegate, Ledezma, first reminded everyone of the colonial present, stating that the "North has environmental debt because for many years they used our natural resources without paying anything in exchange. Biodiversity cannot be assessed in economic terms. For centuries it has been represented in social and cultural terms. It cannot be expressed in monetary terms." She went on to say:

> Our delegation rejects innovative financial mechanisms linked with biodiversity at this time, including the GDM. It must ... be made clear our environment is not for sale. Some would like it to be put on sale saying that we can only save what has value. Now, that is a wrong vision of things. I don't want to expand on capitalism and its nature, we don't want to repeat the same errors that have happened in other conventions.

From the perspective of the Bolivian delegate, the GDM represented a continuation of the North's commodification of the resources of the South and should be resisted on those grounds.

Other opening statements raised concerns about the GDM, concerns that – in spite of the intergovernmental lobbying work of Earthmind – had deepened since Nairobi. Almost all Southern governments that addressed the GDM and the bracketed text asked for it to be deleted, including

Figure 7.8 Photo of action at COP 10 (1). © Anne Peterman 2010 (by permission).

Malaysia, the Africa Group, the Philippines, Bolivia, and Colombia. This kind of consensus foreshadowed a death-knell for the initiative at the COP.

On the second day of the negotiation, a group of NGOs, including Global Forest Coalition, Global Justice Ecology Network, Third World Network, Ecoropa, and Via Campesina, organized a small action with a message similar to that of the ALBA grouping. About 20 people held signs at the entrance to the conference center with the messages: "No to the Greed Economy"' (a spin-off of the "Green Economy"), "Planet Earth is Not for Sale," and "Yes to rights, equity and biodiversity" (see Figure 7.8 and Figure 7.9). These growing debates began to unravel the consensus forged in Nairobi, where Parties had tacitly approved innovative financial mechanisms in principle, even if expressing concern with the specific proposal of the GDM.

Circular debates

Much of the GDM-related debate in Nagoya took place in a small group negotiation that often ran late into the night. Around 30 delegates crammed around a small table in the designated room, with observers sitting around them in a second row. The room was often hot. At the

Figure 7.9   Photo of action at COP 10 (2). © Anne Peterman 2010
(by permission).

front of the room sat an always-on projector with the recent iteration of
the text, in the ubiquitous track changes mode. All conversations in these
smaller group negotiations took place in English, as there is no translation into the UN languages except in the formal sessions. In each session,
delegates went through the decisions paragraph by paragraph, making
comments and suggestions for changed or new text. Elsewhere in the
conference complex, negotiations went on for a new protocol on genetic
resources, a new strategic plan, and many other issues. The purpose of all
these negotiations was to come to consensus on all of these issues, to rid
the text of the brackets and have a clean set of decisions to guide the work
of Party signatories and the international community into the future.

While the co-chairs of the meeting – who were Swiss and
Indian – attempted to keep the clean, "unbracketed" text unbracketed,
some Parties arrived with mandates to dispute the consensus text agreed
upon in Nairobi. This frustrated many. The delegate from EU asked with
irritation, "Why do we even have a WGRI if everything is reopened?"
To this, de Castro-Müller from the Philippines responded with irritation:
"In reference to EU – not all countries can be at all these meetings all the
time." She continued:

> The importance of a truly multilateral event is that it is open, transparent,
> democratic. Unfortunately for many developing countries, they do not
> have big delegations ... and therefore we should give a hearing to those

who are not everywhere, at every time. I think we should not close the argument by saying that there was a small group that met and talked, so we cannot talk about it again.

As is common at negotiations, the debate continued over the weekend, and on Saturday afternoon, halfway through the negotiation, the floodgates opened on the question of innovative financial resources. In this session, Ledezma (from Bolivia) asked for a reference to the outcomes from the People's Agreement – text clearly opposed to market mechanisms – to be inserted in the text of the "innovative finance" decision. This reference, she felt, was critical alongside the already included outcomes and processes such as The Economics of Ecosystems and Biodiversity project, the contentious Berlin workshop on IFMs. In the small, hot room, the response to her suggestion was much snickering and sideways glances. Apparently many countries of the North did not support the Rights of Mother Earth Conference and its People's Agreement. In fact, Northern delegates eventually agreed to delete references to all conferences and initiatives, seemingly just to avoid mentioning the People's Agreement and the Bolivian conference at all.

However, the key contribution from Bolivia was not in insisting on conference references, but in a broader concern with IFMs. As Ledezma explained: "There are new and innovative financial mechanisms, but they could also undermine the objectives of the Convention. We need to examine and evaluate them ... to consider the problems. One problem we have identified is the possibility of commodification of nature." Any new and innovative financial mechanisms, she argued, must adhere to principles, principles that protect against negative aspects they might generate like "the commodification of nature." Ledezma went on to suggest an additional subparagraph to the text, outlining principles that any IFMs must adhere to:

Considering that Innovative Financial Mechanisms could have potential benefits but also possible problems, these mechanisms shall respond to the following principles:

a Consistency with the objectives of the Convention,
b Take fully into account the different circumstances of developing countries,
c Take fully into account the intrinsic value of biological diversity and of the ecological, genetic, social, economic, scientific, educational, cultural, recreational and aesthetic values of biological diversity and its components,
d Shall not allow market mechanisms,

e Recognize and defend the rights of nature and to ensure the full respect of human rights, including the inherent rights of indigenous and local communities,

f Ensure net benefit for biodiversity, thus not permitting that improvements in one country compensate for worsening situations in other countries.

(Co-chairs' non paper, October 22, 2010, available from author).[25]

The Bolivian subparagraph clearly set out a defense of the rights of Southern countries, Indigenous Peoples, and local communities, and outlined a strong opposition to the use of market mechanisms in international biodiversity conservation.

In addition to these principles, Bolivia noted that any IFMs must also have safeguards to protect against their potential negative aspects. The safeguards suggested are similar to the principles:

a They must not provoke financial speculation,

b They must not provoke commodification of nature,

c They must not result from actions that could undermine the achieving of the Convention's three objectives,

d The rights of indigenous peoples and local communities are incorporated, including their full and effective participation,

e They must not provoke any additional burden for developing country Parties.

(Co-chairs' non paper October 22, 2010, available from author)

These principles and safeguards were clearly problematic for many (Northern) countries, especially in their overt opposition to market mechanisms.

At first, several countries from the South supported Bolivia. Brazil stated overtly that principles, in general, were a good idea. The Brazilian delegate noted:

We are dealing with a financial crisis, because the market lacked the ability with the invisible hand to look at other principles than profit and arbitrage, and adding value on things that do not have value. And I think we are giving a very good message if we can not only welcome IFMs, but also give principles to that activity.

The Philippines' de Castro-Müller also agreed that IFMs might be important, noting that "we are not against innovative financial mechanisms" but "we must make sure that there is no undue damage." The Philippines went on to suggest a compromise text, adding a paragraph focused on the need to ensure that any new mechanisms would not impact negatively on countries in the South, noting "the fact that economic and social development and eradication of poverty are the first and overriding priorities of the developing country Parties" (Co-chairs' non paper October 22, 2010; available from author).

The response from several countries in the North was strong. For example, the Swiss delegate focused on what he saw as a contradiction between text that rejects market mechanisms and embraces the idea of IFMs, which he said is like "barbecuing snowballs." The discussion of principles and safeguards (and the Bolivian interventions in general) irritated many around the table. Another Swiss delegate noted that this approach in the text would send "the wrong kind of signals to the private sector." New Zealand noted that the wording, referring to the "commodification" language, was "ideological ... and could lead to unhelpful discussions." Australia proposed compromise text by simply noting that any IFMs must be consistent with the objectives of the Convention. Many countries from the North supported Australia's suggestion.

Despite the irritation with Bolivia, there was general agreement, even among "developed" countries, that both principles and safeguards were a good idea. This agreement on the need for safeguards in many ways signals a recognition that the market mechanisms meant to "internalize the externalities" (like carbon emissions and biodiversity loss) of global economic processes in fact create additional externalities that need to be addressed, like land dispossession. As Lohmann (2011, 113) notes, in environmental market-making, "any process of internalization, in short, creates its own externalities."

But concerns about future potential negative externalities were overshadowed by a larger question: Were environmental markets working at all? Was the carbon market really helping address runaway climate change? The Bolivian protest of market mechanisms at COP 10 was propelled not only by ideological beliefs about the rights of nature, but also by evidence that the existing international climate policies were not working, particularly the use of offsets. The People's Agreement calls for a new phase of the Kyoto Protocol (from 2013 to 2017), "under which developed countries must agree to significant domestic emissions reductions of at least 50% based on 1990 levels, excluding carbon markets or other offset mechanisms that mask the failure of actual reductions in greenhouse gas emissions." In addition to its failure to reduce emissions, delegates noted, the CDM has failed to provide development finance evenly. The CDM is sometimes called the "China Development Mechanism," referring to the large amount of emissions credits that go to a small group of countries and in particular to China, bypassing most nations. Thus, opposition to market mechanisms at COP 10 was also based in delegates' experience with the international carbon market, a market that has so far done little to advance even the most watered-down version of sustainable development in most countries of the Global South.

No compromise on IFMs

Although delegates worked almost nonstop for 12 days, the negotiation went late into the night on the final day. Friday evening, the Parties broke from negotiations for a couple of hours; a big celebration was planned weeks in advance to mark what should have been the closure of COP 10, and to fete the selection of India to host the 11th Conference of the Parties (COP 11). Wine and beer flowed freely, and delegates enjoyed a colorful display of Indian dance and food. However, the celebration was premature: the final plenary had yet to even begin. The final plenary meeting usually just involves "gavelling through" the decisions that were forged in the smaller groups. At COP 10, though, Twitter feeds with the hash tag "#COP10" fluttered with rumours that Bolivia or Cuba was going to block the passing of the protocol on genetic resources – a move that would likely cause COP 10 to fail. And the document on IFMs arrived at the plenary with undecided text. The last few hours of the COP marked an incredibly unusual situation; the final day and plenary is usually largely ceremonial. The TV cameras stood on guard, waiting for the decisions to be announced.

With much diplomatic back-and-forth that at times verged on comical, the protocol on genetic resources was adopted early Saturday morning, although several Parties (Cuba, Bolivia, Venezuela, Ecuador) put on record that "they could not accept a Protocol that failed to meet the minimum requirements of preventing biopiracy" (Third World Network 2010). COP 10 brought into being what are called the "Aichi targets," which aim to at least halve the loss of natural habitats and expand protected areas to 17% of the world's land area by 2020 and increase marine protected areas to 10% of the world's seas. However, despite the ongoing pleas from the South for an increase in public funds flowing North to South in support of biodiversity conservation, the Parties failed to agree on specific quantitative targets for increased financial resources.[26] This failure closely parallels the failure of Northern to make meaningful commitments in the climate talks.[27] What good, asked many delegates, are bold targets and timelines when most countries are not willing to put forward the financial resources to make anything happen?

The decision on IFMs arrived at the final plenary still riddled with brackets. This was highly unusual; documents usually arrive at the final plenary without brackets, ready to be formally adopted. Southern solidarity on IFMs had dissipated. Brazil, which had originally lent some support to Bolivia, moved away, expressing frustration at what it saw as Bolivia's failure to compromise. During the celebration, small groups of delegates gathered to try to hammer out a final last-minute agreement on IFMs. Norway tried, with allies from Sweden, to salvage some kind of decision. Their compromise proposal focused on developing agreement on a process that would set out principles and safeguards for innovative financial mechanisms. Their proposal also called for the Secretariat to undertake a study on the potential positive and negative impacts of IFMs, including those that might incentivise the "transformation of nature into commodities." Despite these efforts, for the first time in CBD negotiation history, not even a watered-down decision passed. No decision was taken at all, and the carefully discussed and debated text simply disappeared.

Conclusions

Over the course of the 2000s, the global biodiversity conservation community turned its attention toward new strategies for resource mobilization. Between 2000 and 2010, global experts worked to turn global biodiversity loss into a problem amenable to a technical solution of innovative financial mechanisms that could unleash value by bringing entities "inside" the market, thereby turning "dead assets" into "live" ones. The idea that this kind of marketization could simultaneously address both uneven development and environmental degradation is a persistent idea within the popular development paradigm of "green developmentalism," one that continues today. In the case of new biodiversity-related IFMs, the promise was that a new, technical approach could solve geopolitical gridlocks over conservation financing and bring millions of dollars into the project of conservation and the generation of sustainable livelihoods.

This attempt to "render technical" the problem of biodiversity loss, a problem that is deeply historical and shaped in large part by ongoing and vastly inequitable terms of trade and geopolitical power, has the effect of creating a demand for exactly the kind of policy knowledge that certain Northern experts could supply. "Experts are trained to frame problems in technical terms," argues Tania Li (2007, 7). This, Li writes, "is their job," and indeed "their claim to expertise depends on their capacity to diagnose problems in ways that match the kinds of solutions that fall within their repertoire" (2007, 7; see also Ferguson 1994). The result was a fast-growing network of people committed to new mechanisms, policy solutions that could bypass the thornier political-economic and historical definitions of the problem of biodiversity loss. Many people sprang into action, traveling the world to try to convince government representatives and other key figures that the creation of a new financial mechanism could generate the financial resources that conservation so desperately needed. The goal was to obtain political "buy-in" for a new mechanism for conservation finance and to have the official text of the CBD amended to advance this new approach.

The story I've told here is a complex one, involving several individuals, organizations, meetings, working papers, decisions – a wide range of law-and-policy-shaping activities taking place in diverse locations around the globe. Multilateral law and policy is often so complicated as to be impenetrable to outsiders; this impenetrability provides legitimacy to the decision-making, as outsiders cannot see the detailed histories behind the official text. I tried to retain aspects of these detailed histories in order to show the embodied processes through which these policies are advanced and debated. Careful attention to details like side meetings, brackets, and even the body language of delegates elucidates the ongoing tensions over unequal North–South relations that have long shaped global biodiversity politics.

And indeed, the tensions ran high over IFMs. The attempt to redraw the line between "inside" and "outside" the market became a major battleground. Representatives from Southern nations – including Bolivia, the Philippines, and Kenya, for example – continuously challenged the idea of IFMs and the GDM. They argued that the GDM had been created by Northern experts. They drew on experiences with the CDM, a mechanism that had created a very limited flow of resources, mainly to China, and failed to address the emissions problem that was its target. Primarily, however, Southern representatives called attention to history: they noted that past patterns of resource exploitation that had resulted in biodiversity loss had disproportionately benefited the North and impacted the South; they reminded Northern nations of Article 20 of the Convention, of the promise made at Rio that developed countries would help cover the incremental costs of biodiversity conservation. In these ways, the proposal of a technical mechanism actually reproduced – and in some cases amplified – the very tensions that the mechanism was intended to address.

In the end, IFMs and the GDM failed to obtain intergovernmental assent. This failure was not final nail in the coffin of the "biodiversity capital." Even at these meetings, initiatives like the TEEB project and proposals for biodiversity offsets stormed ahead in the corridors (see

MacDonald and Corson 2010). The GDM also marched on, although without its association with the fraught CDM. It was rebranded as the Green Development Initiative and its proponents are still trying to create an internationally approved biodiversity credit.[28] Just months after the failures of innovative finance in Nagoya, conversations about biodiversity offsets and habitat banking started in the UK, upon the premise that such a market is needed to "stump up enough cash to deliver our biodiversity objectives" (Caldecott 2011), setting off a heated debate not dissimilar to the one in Nagoya. And in 2014, big financial actors like JP Morgan and Credit Suisse began to join the choir of voices calling for private, return-generating capital to fill giant gaps left by austere governments – North and South (e.g. Credit Suisse et al. 2014, NatureVest 2014).

And so, the promise of the "big big money" lying just outside the market remains seductive. The foundational story remains strong: enormous amounts of conservation capital sit just around the corner, awaiting dead (but living) assets to become truly alive.

Notes

1   Article 20 states that Parties in the North must pay for the incremental costs of conserving biodiversity in the Global South: "The developed country Parties shall provide new and additional financial resources to enable developing country Parties to meet the agreed full incremental costs to them of implementing measures which fulfill the obligations of this Convention." All articles of the CBD can be accessed at http://www.cbd.int/convention/text (last accessed February 26, 2016).

2   Most notably, the Clean Development Mechanism – the carbon trading mechanism under the Kyoto Protocol – became operational in 2006, and many saw the CDM as a model on which a mechanism for biodiversity could be loosely based. As well, many actors in global biodiversity politics were showing greater interest in the role of finance and business in conservation, seeking out opportunities to involve the private sector (see chapters 5 and 6; see also Brockington et al. (2008), MacDonald (2010).

3   See decision VIII/13 paragraph 4 for the strategy on resource mobilization. Also in Curitiba, Parties were encouraged to explore "options for innovative international finance mechanisms to support the programme of work on protected areas" (Decision VIII/24, paragraph 18 f). Available at http://www.cbd.int/decisions/cop/?m=cop-08 (last accessed February 18, 2015).

4   See Decision IX/11. Available at http://www.cbd.int/decisions/cop/?m=cop-09 (last accessed February 18, 2015).

5   The Ad Hoc Technical Expert Groups (AHTEGs), composed of nominated experts who provide reports and recommendations, are one example

of a sub-group that participates in smaller, intercessional negotiations. There are also Party-based negotiations that act like "mini-COPs." These include meetings of the Subsidiary Body on Scientific, Technical and Technological Advice (SBSTTA), a body that consists of scientific and technical experts from the Parties. The SBSTTA gives scientific and technical recommendations to the COP and serves as a first pass for identifying contentious issues and sticking points in the text.

6   However, it is important to note that representation is shifting as governments from the North cut back on delegations and countries like China and Brazil begin to send more delegates.

7   The negotiation in Nairobi took place immediately after what Northern countries were saying was the largest ever replenishment of the Global Environment Facility (GEF), the fiscal mechanism for Articles 20 and 21: $4.25 billion over four years. Countries from the South claim that this is not enough, only a drop in the bucket. They also note that much GEF financing must now be leveraged with other resources, resources that usually are borrowed, resulting in greater domestic debt.

8   The "Policy Options Concerning Innovative Financial Mechanisms" document can be found at http://www.cbd.int/doc/meetings/wgri/wgri-03/official/wgri-03-08-en.doc. The following quotes all come from this document (CBD 2010c), last accessed February 18, 2015.

9   The entire document is 18 pages long, but not all of this is negotiated text. Only the final two pages are up for debate and discussion because these form the basis of the legal decisions to be made. During discussions, areas of disagreement are put into square brackets, but participants seek to minimize these brackets in order to make the work of the COP smoother.

10   When impasse hit on Thursday of that week, the chair also established what is known as a "Friends of the Chair," which is an exclusive negotiating space. The group included South Africa, Brazil, India, Switzerland, the EU, and Canada, and met late into the night.

11   Iran suggested changing the paragraph and referencing the outcomes of the workshop only in the preamble.

12   Recommendation 3/9. Available at http://www.cbd.int/wgri3/meeting/Documents.shtml, last accessed February 18, 2015.

13   Francis Vorhies was the first economist at IUCN and a key actor in drawing IUCN together with business.

14   A background paper was commissioned for this event at COP 9. It is available at http://gdm.earthmind.net/files/balancing-biodiversity-Final-Draft_070208.pdf (last accessed February 19, 2015).

15   Document available from author.

16   A report commissioned by one of the same ministries (Housing, Spatial Planning and Environment) in December 2008 demonstrates this interest, and sets up the push for the GDM. The report, titled *Economic Instruments for Biodiversity: Setting up a Trading System for Europe,* begins the process of addressing the impact that Europeans have on biodiversity by "attaching economic value to biodiversity" (Blom et al. 2008, 1). The

report lays out a "Biodiversity Trading System" that parallels the European Trading System for carbon in the EU. The report even tries to develop a methodology for assessing how European goods and services are attached to biodiversity loss.

17　Swanson is a familiar figure in global biodiversity policies and is especially concerned with transforming biodiversity from an entity external to the economy to one that is internal. He created the supply-and-demand curve for biodiversity discussed in chapter 3 and has advocated a kind of "unleashing" of biodiversity capital through intellectual property systems for genetic resources; he promotes bioprospecting as a "green development" option.

18　The workshop included 11 governmental representatives; nine of these came from North. Also present were 10 NGOs (eight of which were Northern-based). Seven international organizations participated, including representatives from the International Finance Corporation, the Secretariat to the Convention on Biological Diversity, the OECD, UNEP, TEEB, and UNDP. Six business representatives attended, representing the following organizations: the World Business Council on Sustainable Development, Greenpalm, Royal Dutch Shell, and Wilmar International. Four academics (from the University College London, Cambridge, Universidade Federal do Rio de Janeiro, and Duke University) were also present. Pavan Sukhdev, the head of the TEEB initiative (now head of UNEP Green Economy Initiative), played a role, facilitating working meetings.

19　Like the CDM, the GDM would require governments to agree to some level of constraint, to monitor land conversion, and "to recognize the contractual obligations to conserve land for biodiversity in return for the sale of biodiversity credits" (Mullan and Swanson 2009b, 19). Such an agreement by government, proponents say, should ideally take place through the Convention on Biological Diversity process, just as the CDM found expression in the United Nations Framework Convention on Climate Change (UNFCCC) process. The GDM would also require private sector participation as businesses are envisioned as the main participants in the trading system as buyers and potentially sellers. And just as defining a "Certified Emission Reduction" under the CDM requires meeting a series of scientific, technical, and economic criteria, so too the GDM "will require methods for measuring changes in biodiversity that can be applied to multiple ecosystem types in varying condition" (Mullan and Swanson 2009a, n.p.).

20　A website was created – http://gdm.earthmind.net/(now http://earthmind. org/gdm/) – which chronicled the policy's advance, providing a rich repository of policy history, including many PowerPoint presentations.

21　G77 is a loose coalition of developing countries that sometimes work together at UN negotiations. It started with 77 member countries but has since grown to over 130.

22　In order to access funds from the GEF, which are intended to cover the "incremental costs" of biodiversity conservation (among other environmental improvements) in the developing world, applicant countries must meet certain co-financing arrangements. However, to meet these arrangements, developing countries need funds, most of which come in the form of loans and therefore exacerbate debt load. As de Castro-Müller explained: "Co-financing is not leveraging, it is a problem ... Before you get GEF monies you need to first get monies elsewhere, and these are mostly loans. That means that co-financing is putting developing countries more and more in debt."

23   ALBA stands for the "Bolivarian Alliance for the Peoples of Our America." The 11 member nations are Antigua and Barbuda, Bolivia, Cuba, Dominica, Ecuador, Grenada, Nicaragua, Saint Kitts and Nevis, Saint Lucia, Saint Vincent and the Grenadines, and Venezuela. The organization is associated with socialist and social democratic governments and is an attempt at regional economic integration based on a vision of social welfare, bartering, and mutual economic aid, rather than on trade liberalization as with free trade agreements.

24   See https://pwccc.wordpress.com/support/ (last accessed February 19, 2015).

25   A non paper refers to informal negotiating text circulated amongst delegations for discussion.

26   On the final night in Working Group II, Brazil requested that "by 2020 at the latest, to increase substantially from current levels, the mobilization of financial resources for implementing the strategic plan, 2011–2020, from all sources and through a consolidated and agreed process, reaching at least 200 billion dollars." Donor countries opposing targets used a lack of understanding about how many resources were actually needed to implement CBD decisions, and missing baselines and measurement methodologies, as an excuse to commit to any specific amounts.

27   However, the final COP 10 decision involved a number of intercessional research and activities in order to set a target on financial resources at COP11 in 2012; see COP/DEC/X/3 para 8 (i), available at https://www.cbd.int/decision/cop/default.shtml?id=12269 (last accessed April 28, 2016).

28   See http://gdi.earthmind.net (last accessed February 19, 2015).

8

The Tragedy of Liberal Environmentalism

"So There Is Lots of Stuff Going On … But When You Think About It, Very Little Is Going On"

Will the twenty-first century be the century in which nonhumans are recognized for their enterprising nature, for being as indispensable as Walmart? Will it be the century that biodiversity finally takes up its rightful place *inside* the market, where it can unleash billions of capital for "green development"? Given the broader political-economic context, the answer, perhaps intuitively, is yes. Enterprising nature is an approach that fits well with global business as usual.

This book focuses on an era in which the discourse and practices of enterprising nature were proliferating rapidly in global biodiversity politics. And the enterprising continues. For example, in 2012, 39 financial institutions signed the Natural Capital Declaration (NCD). The NCD aims to create a standardized internationally agreed framework to be used by companies and financial institutions to account for and manage natural capital. Signatories include large global financial institutions like Rabobank, Standard Chartered, and National Australia Bank. In October 2015, President Obama directed all federal agencies to incorporate the value of ecosystem services, or "green infrastructure," into their planning and decision-making. New rounds of "conservation finance" conferences are bringing together financiers, project developers and NGOs – such events now take place at the New York city offices of Credit Suisse and involve the world's largest financial institutions, such as JP Morgan and Goldman Sachs. The proliferation of such initiatives,

Enterprising Nature: Economics, Markets, and Finance in Global Biodiversity Politics, First Edition. Jessica Dempsey.
© 2016 John Wiley & Sons, Ltd. Published 2016 by John Wiley & Sons, Ltd.

involving high-powered geopolitical and financial actors, appears to suggest that enterprising nature is on its way to being mainstreamed in global capitalism.

Yet international professionals, elites, and financiers have been trying to "sell nature to save it" (McAfee 1999) for a long time, and are still trying to figure out what they need to do to turn biological diversity into a legitimate economic actor that can save its own life. Global biodiversity experts struggle to define the unit they want to save through its commodification (chapter 6); proposals for "innovative financial mechanisms" fail to receive intergovernmental assent (chapter 7). Even mundane initiatives in accounting face difficulties, and experts are asked to provide ever simpler forms of ecological knowledge that can be "relevant" and thus capable of being incorporated into firm and government decision-making (chapters 4, 5). Ecological-economic calculative devices do produce new "facts of life," but these do not readily lead to reformatted political-economic relations. The most status quo–affirming and supposedly pragmatic approach to "saving the planet" is not smooth or easy; it is better conceived as Sisyphean.

Enterprising nature exists in an entirely paradoxical situation. It is at once a totalizing mainstream discourse and one that exists on the margins of political-economic life, on the outside of many flows of goods, commodities, and state policies. Over the past few years, studies on "conservation finance" produced by organizations such as The Nature Conservancy, JP Morgan, and Credit Suisse reflect the findings of this book. They declare the coming exponential growth of enterprising natures while reiterating persistent challenges: for-profit, private sector investment in conservation remains very limited, and most financing still comes from conventional, well-established channels of domestic government funding, development assistance, and philanthropy.[1] Once considered the cash-flush messiah for tropical forest conservation, the forest carbon market faces a problem of over-supplied credits and low prices (Global Canopy Project et al. 2014). The global forest carbon market transacted a paltry 216 million dollars in 2012 and 192 million dollars in 2013, amounts similar in size to the sales of a single Walmart store.[2] We may have hit peak carbon market well in advance of reaching anything close to peak oil.

In short, biodiversity markets remain small – marginal even in the world of conservation finance, infinitesimal in the world of capital flows writ large (Dempsey and Suarez 2016). Instead of picturing "liquid biodiversity capital" zooming across the globe in smooth corporate jets guided by slick capitalists, I suggest we imagine enterprising nature as clunky and plodding, something more like a jalopy puttering along with flat tires and occasional backfires (but with a professional pit crew working furiously on a project of constant reassembly).

I don't want to give the impression that nothing is going on – the jalopy is still moving. New financial products are emerging with support from the biggest financial institutions in the world.[3] New manifestations of forest-backed bonds are under discussion. New hybrid institutions are emerging, like NatureVest – a collaboration between The Nature Conservancy and JP Morgan. The conservation finance conference in 2016 held at the offices of Credit Suisse in New York City was filled with participants in much fancier suits, representing truly gargantuan financial firms; these events were much more high powered and professionalized than the 2008–2009 conferences described in chapter 6. Yet the 2016 conference asked a question similar to the ones being asked at events seven years earlier: how can we scale up return-generating biodiversity conservation? The participants still spoke of challenges outlined in chapter 6: deals take forever to close, few investors understand the work firms are doing, there is too little transparency, more professionalization is needed. Much of what is going on is heavily supported by the capital of high-net-worth families and philanthropic and public institutions that are willing to take on more risk than mainstream investors.

While the jalopy bounces along, I hold to my conclusion: enterprising nature is a dominant story about how to change the world, but it remains marginal in practice. Conceptually dominant, but substantively marginal. As one investment banker said at the 2009 New York conference about so-called biodiversity markets: "there is lots of stuff going on ... but when you think about it, very little is going on."

The Radical Project of Enterprising Nature?

This book traces the rise of a new mantra in conservation: "to make live, one must make economic." But the book also traces this mantra's persistent marginality. People who seem so powerful and influential, individuals like Gretchen Daily and Walter Reid, who reside in prestigious institutions like Stanford, who are invited to address heads of state and to advise major multinational initiatives, are situated simultaneously outside and inside mainstream institutions and political economic power relations. They are not CEOs of major corporations, nor leaders of governments. Rather, they are trying to move those people and those institutions; they are trying to convince governments to adopt full-cost national accounting and to consider time frames beyond the next election cycle. These tasks are hardly easy, even if all the best ecological-economic evidence underscores their benefits.

My research is dogged by the question of how a conservation approach that is so in line with mainstream political-economic logics can be so difficult to implement, even in a watered-down, pragmatic form. Given that the idea of internalizing externalities is so consistent with orthodox economic thinking, why aren't national governments creating regulatory frameworks to this end? Why is it so hard to build the conditions to realize the green economy, to finally bring biodiversity inside market relations (and maybe even save it)?

The project of making enterprising nature – despite being the most politically palatable approach – faces many challenges. Biological diversity is, as I have shown throughout this book, enormously difficult to domesticate into a quantitative form; it is difficult "to enterprise." Biodiversity – life on earth – is hard even to count, never mind to parse in terms of ecological functions and services that can then be priced or monetized. Finding these connections is a research project in ecology that is at least a half century old, one that seems to move forward while simultaneously opening further unknowns and uncertainties about ecological relationships, especially in the context of a rapidly changing climate.[4]

And even if one accepts "imperfect proxies" or abstractions that simplify all the complexities and render unknowns into probabilities, actually transforming state accounting or firm risk assessment to account for biological diversity remains daunting. Internalizing externalities, it turns out, poses impressive challenges to the status quo. Decisions – say, to build new energy infrastructure to extract fossil fuels (i.e. pipelines, refineries), or the continued destruction of mangroves for shrimp farms – are not likely to be reversed due to new calculative figures about the "full costs" of ocean acidification and mangrove destruction. While economically "stupid" or "irrational" decisions may haunt us in the future, those stupid economic decisions pay in the present.

Earlier in the book I mentioned a revealing study commissioned by The Economics of Ecosystems and Biodiversity project (TEEB). Tallying up the total "unpriced natural capital" (ecological materials and services that businesses currently do not pay for, such as clean water and a stable atmosphere), the study found that none of the globe's biggest businesses would be profitable if it had to pay for those services (Trucost 2013). This fact illustrates a very large and intractable problem: profit and power structures in the global political economy depend deeply on these externalizations, and efforts to alter externalizations mean confronting these formidable forces. Enterprising nature must still fight battles with the axes of power and profit in the worlds of agribusiness, oil and gas, and extractives (to name some of many), and the governments that are tied to the resource rents from them. Convincing decision-makers to internalize the full cost of goods and services produced and provided by nature is like trying to get a Goldman Sachs executive to give up his obscenely high bonus – in short, incredibly difficult. What this suggests is that attempts to create enterprising nature, as the most economistic, business-as-usual approach to solving the sixth extinction, are in part foiled by economic self-interest and by contemporary concentrations of wealth and political power.

This is the tragic story of enterprising nature. As supposedly pragmatic and neoliberal as it is, as much as it reflects social norms, purporting to smoothly lead the way into the known future where economy and environment can co-exist in perfect harmony, it is still in many ways too radical and too challenging to the status quo to become mainstream. Attempts to enterprise nature are simultaneously paradigmatic of neoliberal environmentalism *and* threatening to the foundational characteristics of contemporary capitalist social relations. They are threatening in that they pose challenges to what socialist ecofeminist Maria Mies (1986, 1998) calls the "iceberg of capitalist accumulation," wherein profit-making sits not only on the visible exploitation of wage labor (above the water line) but also upon layers of exploitation under the water line, including the unpaid work of nature (see also Fraser 2014). Jason Moore (2015) terms all this unpriced work "cheap nature" and, like Mies, argues that this cheapness is a "fundamental condition of capitalist accumulation" (Moore 2015, 2). Analyzing Mies's and Moore's arguments in relation to this study alters the way we understand the invisibility of biodiversity in economic processes. Biodiversity loss is not simply an unfortunate side effect that can be fixed through accounting or market-making; rather, such loss might be thought of as critical to the functioning and stability of capitalism as we know it. Enterprising nature exists as both a hegemonic approach to nature *and* one destined to continue in the form of briefly illuminating "fireflies," to use the words of one financial executive from the 2009 New York conference: that is, short, quick bursts of light in the dark night.

This arrested development of a pragmatic, neoliberal-aligned environmentalism does not mean that these "fireflies" are inconsequential or entirely benign, and these effects need to be studied carefully in situ. All development projects have winners and losers, social divisions created or deepened along fault lines of race, class, geography, and gender. In situ, enterprising nature produces new dispossessions (e.g. Cavanagh and Benjaminsen 2014) and results in hybrid flows of state and private capital (e.g. McAfee and Shapiro 2010). Enterprising nature can also, with enormous political effort on behalf of rural social movements, affirm the value and necessity of *campesino* environmental stewardship (Shapiro-Garza 2013). This complex and crucial on-the-ground research shows that the social effects of enterprising nature are not wholly predictable or consistent, but are indisputably real for many different people.

What I am saying here is not incompatible with the findings of these studies, but my focus on the global circuits of power and knowledge illuminates other corners of contemporary biodiversity politics. In this book, I argue that the story of enterprising nature illustrates the tragedy of liberal environmentalism almost 25 years after the Rio Earth Summit. Liberal environmentalism encompasses the classic compromise and pragmatic stance of sustainable development, an approach that aims to make environmental concerns compatible with economic growth within predominantly capitalist markets and states, a compatibility to be achieved via heavy doses of science and technology (Bernstein 2002). But liberal environmentalism runs deeper than this. It is an approach premised on an idea of a smooth space of politics, one where all the different players can find common ground through dialogue or, even better, through the purportedly neutral signifiers of numbers and money – liberation by calculation. It aims, as much as possible, to avoid dirty, asymmetrical, bloody politics; it lives in a world that James Ferguson aptly terms an "anti-politics." Despite its marginality, the enterprising nature story is a powerful salve, a kind of chicken soup for the environmentalist soul. It is a story that manages, moderates, and mediates the problem of the fraying web of life with its message of ever-increasing rational decisions, its story of ever-improving governance and progress at the hands of the right ecological-economic facts.

As with the rise of bioprospecting and biodiversity in the late 1980s, however, we remain in a kind of liberal environmental "waiting room."[5] The destination is known, demonstrated by the illustration on the cover of this book. We are en route to the coming bioeconomy or, perhaps, to a coming global bio-political-economy where all social, economic, and natural values can be accounted for within a single analytical system, aligning global socioecological needs, national interests, and economic growth. Yet arrival is always just out of reach, just past the next problem, just over the next epistemological or policy hump. Rather than wait or keep reaching, it is time to change the strategy and open up space for more political narratives, for other end points and strategies.

Finding a New Pragmatic Politics

At the start of the book, I charted what many saw as the failures of conservation. People don't care about nature for nature's sake, I heard over and over. Conservation is too focused on ecosystems as if they didn't include humans, others said. These are legitimate concerns. Many in the field of biodiversity conservation have turned toward enterprising nature as a singularly pragmatic approach – as our only hope on a planet filled with disconnected people who can only understand dollars and cents. From the vantage point of this book, however, enterprising nature doesn't seem overly pragmatic, at least not in its current form. I don't suggest we flee into theory or experimentation, though. Rather, a new sensible and even practical politics is not only possible, it already exists.

Between scarcity and abundance

As someone reared on Western environmentalism's constant talk of limits, of constraint rather than opportunities for flourishing, I was at first caught off guard by Anishinaabe writer Leanne Simpson's words. Outlining the Anishinaabeg concept of *mino bimaadiziwin*, she writes: "The purpose of life is this continuous rebirth, it's to promote more life. In Anishinaabeg society, our economic systems, our education systems, our systems of governance, and our political systems were designed with that basic tenet at their core." A key consideration for her nation is "how much you can give up to promote more life." This idea is strikingly dissimilar to the narratives in environmental policy and science that talk about rationing, optimizing, and managing scarce resources. To even speak of abundant ontologies feels almost risky for people (such as me) who have long feared environmental overexploitation and its cliff edge of catastrophe; we might be concerned about a return to a false cornucopianism. But Simpson is not saying that there are no biophysical limits; rather, she is pointing to actually existing ontologies and socio-ecological epistemologies rooted in visions and practices of proliferating life.

Writer James MacKinnon shares with Simpson the aim of abundance and liveliness in socioecological relations. His book *The Once and Future World* begins with stories of lost abundance, of jungles emptied by bush meat hunters, the fading of British tree sparrows, the vanishing of the Chinese river dolphin, Caribbean reefs that host at least two tonnes *less* fish per hectare than in the seventeenth century. Reflecting on a 1902 sketch of a fisherman spearing a rock cod from a boat in the Fraser River near Vancouver, BC, MacKinnon writes that such a feat would require "a sea so jim-jammed with life that it beggars belief" (MacKinnon 2010). "Our natural world," he says, "is a fraction of what it was before the mass culls and oil spills of the human era." MacKinnon describes our current Earth as a "10% world," by which he means that we inhabit a planet that has only 10% of the natural variety and abundance it once did. For MacKinnon, this is not a call for "some romantic return to a pre-human Eden." Rather he posits that "a story of loss is not always and only a lament; it can also be a measure of possibility. What once was may be again." MacKinnon is not talking about more parks and protected areas, or wilderness, but rather setting his vision higher, writing of an "Age of Restoration" and an "Age of Integration," within which "human beings can learn to live not only alongside but also among more species, in more abundance, than we ever have before." This aim is not nostalgia motivated by "wilderness lost," but forward-looking hopefulness, guided by a desire for lively abundance and co-habitation, a desire to inhabit rich socioecological worlds.

What Simpson and MacKinnon present are the sketches of what we might strive for in biodiversity conservation: an end goal that is rooted in abundance, not in rationing or optimization. This is a different narrative than that of "nature as Walmart," and one that requires imagination beyond cap and trade, or economic-ecological modeling of "trade-offs." An argument that starts from a desire for abundance in nonhuman life forms – a desire for more kinds of bodies, in greater numbers – and for abundant and diverse ways of living between humans and nonhumans is like heresy in the rationing, austerity-focused discourse that swirls around us. It's heresy to some fundamental narratives of scarcity in economics, biology, and environmentalism (Haraway 1991).

I believe that many of the trustees I describe in this study would agree with Simpson and MacKinnon. Many, I think, would agree wholeheartedly with their vision, although perhaps simultaneously noting that it is "pie in the sky" thinking, idealistic and unrealistic. Meanwhile, though, I have argued that the "will to enterprise" seems unrealistic and untenable; enterprising nature exists on an ever-receding horizon. This is what we learn from 25 years of promissory visions to bring diverse life forms inside market and economistic calculations; this is what we learn from attempts to create the conditions for nature to pay its own way. While trustees in the circuits of my study may cry that "we have tried the intrinsic value approach, and failed," I would respond that we had the wrong target in the first place; we brought together the wrong ideas, the wrong actors, and perhaps the circuits of power and knowledge must be otherwise.

Troublers of liberal environmentalism: Towards biodiversity justice

Rosemary Collard, Juanita Sundberg, and I take up MacKinnon in conversation with decolonization movements like Idle No More and social movements like Via Campesina in our "Abundant Futures Manifesto" (2014). We argue that while conservation should not be organized around the colonial myth of Edenic natures past, it must necessarily continue to look to the past. Conservation must examine histories not only to see what could be in terms of nonhuman abundance but also to understand how we arrived at where we are today, in a world of social inequities and ecological impoverishment. Twenty-first century international conservation needs to reckon not simply with "poverty reduction," but with the ongoing ruination wrought by colonialism and capitalism, by structures and processes that erase distinct ways of living and being. It needs to move away from notions of universal, unilateral value determined from above, whether from colonial administrations or models from afar. We are certainly not the first to make this point. And we don't have to look very far to find people and organizations that are already enacting such a politics. Global biodiversity politics has many troublers of the smooth space of liberal environmentalism.

Take, for example, the International Collective in Support of Fishworkers (ICSF), an organization based in India that works closely with the World Forum on Fisher People and other local fisherpeoples throughout the world. Over email and in negotiations, Chandrika Sharma from the collective would remind scientists, bureaucrats, and other NGOs that local fisherpeople must not bear the brunt of policies to conserve marine biodiversity. Chandrika was willing to engage in conversations and dialogue with others on how to solve these problems, on how to place the lives and knowledge of fisherpeople within policies seeking to address marine habitat loss and species extinctions. Her position was not anti-biodiversity or anti-wild, but she insisted that experts and bureaucrats account for their god's eye view and confront the political-economic realities that lead to the decline of marine biodiversity, particularly industrial fisheries. Chandrika died suddenly in 2014; she was aboard the Malaysian Airlines plane that went missing. Still, ICSF remains one of many troublers of the god's eye view in global biodiversity politics, questioning the location from which conservable natures are defined and demanding that these knowledges and institutions be accountable for their real and potential effects.

There are also what we might call the troublers of austerity: people, groups, and even countries questioning the ever-present story of scarce monetary resources. Malaysian-based Third World Network (TWN), for example, is an organization that continuously calls attention to the fact that the responsibilities for global environmental problems are highly lopsided. Around global biodiversity meetings and negotiations, the ever-fierce, whip-smart, and funny Chee Yoke Ling from TWN continues to bring histories of uneven development and unequal terms of trade into biodiversity negotiations, into multilateral conversations all over the world. Troublers of austerity, at their best, go beyond demands for monetary resources in the form of aid, focusing also on the need for political responses, for changes to the very make-up of global capitalism. Chee Yoke Ling states that "profound economic transformations" are necessary, meaning significant reforms to "global trade, investment and financial rules and architecture" that can remove the "structural obstacles to sustainable development."[6]

While attempts to enterprise nature focus on making existing powerful people and institutions see biological diversity, there are movements and organizations who focus on creating political power. Those who we might call "troublers of scarce power" refuse to believe that power exists only in elite containers; they defy expectations by exploding out of what appears to be a marginal position. This kind of bursting out can be achieved through strategic thinking and huge doses of diligent organizing and solidarity building. Sometimes these initiatives succeed in big ways, as I saw in Curitiba when hundreds of farmers and landless people, working with NGOs and governments, achieved a ban on terminator technologies.[7] I have seen this power-generating force at international negotiations in subtler ways, too, such as when people made short interventions in negotiations by speaking plainly about the effect of biofuel subsidies in Europe and their impact on rainforests and communities, when they playfully awarded the world's worst biodiversity offenders in the hallways, and when they called out in side events my home government, Canada, for unsustainable logging and for the lack of respect for Indigenous rights and title. Each of these moments disrupted the smooth space of liberal environmentalism, they created little zinging jolts of political power, they made people uncomfortable, and they changed the terms of the conversation.

The question, then, is how to ignite a bigger explosion with a chance of slowing biodiversity loss. I don't pretend to have any clear cut or singular answer, but it does seem that there are no shortcuts to deal with biodiversity loss. To say that there are no shortcuts does not mean that there is no role for ecological economics, or valuation, or even natural capital accounting. A price tag on a particular ecosystem service, one that can bring government revenue, could be a powerful tool for a community fighting yet another development on their lands and territories, and seeking alternatives. More sophisticated ecological-economic models might lead to public investments in "green infrastructure." What I mean by no shortcuts is that one cannot avoid battles that need to be fought to insist governments take the long-term vision, to stop bad developments because they do not benefit anyone but private firms or distant countries and consumers. And if we are going to fight battles, it seems to me we should not aim for the most watered down, pragmatic form of conservation, but rather set our sights high.

Alliance and solidarity building among many different kinds of people and institutions is one crucial strategy. One question I ask myself is, could the global circuits of biodiversity power and knowledge become *stranger*?

While at one point it might have seemed strange for biologist Gretchen Daily to converse with eminent economist Partha Dasgupta (see chapter 4), and for them to publish articles together, today the ecologist-economist alliance needs to break into new territory. Perhaps Daily could begin to converse with the ever-inspiring Tewolde Egziabher, the tireless Ethiopian advocate for farmers' rights and agricultural diversity who works at the international, national, and local level to create the conditions necessary to both create and sustain biological diversity. For Tewolde (this is what everyone calls him at Convention on Biological Diversity negotiations), these conditions mean refusing capitalist processes of enclosure over land, waters, and living things, including patents on life (Egziabher 2002). What if ecosystem service scientists spent time, not trying to get their heads around neoclassical economics, but engaging with critical scholars such as Donna Haraway or David Harvey to see what reciprocal learning might occur and what new ideas might be generated about ecosystems and power relations, about scientific discourse and history? An alliance that pays attention to such structures of power and profit, as well as to detailed and rigorous ecosystem science, might, for example, put the powerful InVEST model to work in undermining the ongoing pilfering of the planet, a pilfering that lines the pockets of elites (elites that fund programs like the Natural Capital Project). InVEST could very well be a powerful tool to garner political will and citizen awareness as part of an array of tactics to illustrate how elites and corporations continue to dominate the world's ecosystem services, possibly in collaboration with an organization such as Third World Network. Such alliances could help us understand what we need to do to move toward a justice-oriented and reparative full-cost accounting. Perhaps.

Institutionally, the circuits of power and knowledge might involve strange alliances between big international conservation organizations and social movements. Such circuits would counterbalance (or, ideally displace) the growth of partnerships between conservation organizations and the world's biggest multinational corporations. Green NGOs would focus more on collaboration with the world's largest social movements, including, for example, Via Campesina (the world's largest peasant movement) or climate justice movements. The circuits would have Pavan Sukhdev (now the head of the Green Economy Initiative) attending not (or not only) the World Economic Forum (where the world's elites gather), but (also) the World Social Forum, where the world's social movements gather. And what about if – rather than partnering with the world's largest soya or palm oil companies – conservation NGOs worked to develop a campaign of divestment akin to the one taking place around fossil fuels? Such a campaign would start from a place of principled strength, drawing attention to the way large financial and corporate actors, in entangled dances with many governments, have locked us into patterns of biodiversity loss.

Are these alignments, circuits, and new campaigns possible? Yes, although of course they are not easy to imagine. They certainly would require a level of openness by experts and troublers alike. But if one looks at the mathematical and political gymnastics involved in making nature enterprising, such alignments have more potential than we might think. What I am saying is similar to what political ecologist and critical development scholar Arturo Escobar (1998) called for over 20 years ago, writing just after the formation of the Convention on Biological Diversity:

> One would hope nevertheless that in the spaces of encounter and debate provided by the biodiversity network there could be found ways for academics, scientists, NGOs and intellectuals to reflect seriously on, and support, the alternative frameworks that, with a greater or lesser degree of explicitness and sophistication, Third World social movements are crafting (76).

One can look to the climate justice movement for inspiration and leadership, as a growing international solidarity movement that situates a global environmental problem within contexts of racialized, gendered, geographical, and economic injustices. Biodiversity justice is climate justice's conjoined twin, a movement of scientists, activists, academics, farmers, Indigenous people, urban people, and rural people who demand dramatic redistributions of wealth and power in the service of abundant socioecological futures. Growing such a movement will not be perfect, or easy, but it can start from a place of political and ethical might, in justice, rather than liberal compromise. Justice may even be the wrong term, as I was reminded by Brazilian colleague in a Skype discussion when I used the term. Justice, for her, was simply *too human* a concept. An abundant future must continue to wrestle with how other-than-humans can have wild lives, where they too can live as "uncolonized others" (Plumwood 1993).

. . .

Such an understanding of biodiversity loss – deeply rooted in colonialism, in geography, in power, in class, in race, in gender dynamics – does not lead to simple answers. But I argue that we should not ignore or seek to circumvent difficult questions. There is no shortcut to the messy politics needed to deal with the problem of the monoculturing of life on earth; there is no easy way to confront questions about the human place in nature on a planet of deeply etched asymmetries. I sense there is much appetite and possibility for a "will to abundance" in the circuits of global biodiversity politics. In my travels across and inside the circuits of these politics, I see that enterprising nature is an open, tenuous, and marginal project. There are many points of intervention and lines of struggle, and along those lines there are nodes where it seems possible to change direction, to carve out new lines, or, even better, to join existing ones that are not only more ethical and just, but are also going in a direction that seems more likely to succeed in manifesting abundant futures.

Notes

1   See, for example, Conservation Finance Alliance (2014), Huwyler and Tobin (2014), Madsen et al. (2011), NatureVest and EKO Asset Management Partners (2014), Parker et al. (2012).
2   Since completing this book, I have undertaken research on the size, scope, and character of "for-profit conservation capital," capital that aims to achieve both conservation and accumulation returns for investors. Despite exploding rhetoric around environmental markets over the last two decades, my collaborator Daniel Suarez and I find that the capital flowing into market-based conservation remains small, illiquid, geographically constrained, and typically seeks little to no profit. It is underperforming as both a site of accumulation and as a conservation financing strategy (see Dempsey and Suarez 2016).
3   For example, Althelia Ecosphere is a €105 million closed-end fund launched in 2011 and due to mature in 2021. It invests in agroforestry and sustainable land use and claims that its returns are market rate. Investors are mostly quasi-public institutions like the European Investment Bank, the Dutch development bank FMO, FinnFund in Finland, and the Church of Sweden, as well as the David and Lucile Packard Foundation. In addition to these investors, in 2015 the Fund and Credit Suisse issued "Nature Conservation Notes," debt instruments that generated €15 million of finance from non-institutional investors. Althelia's latest investment is a €7 million commitment toward protecting 570 000 hectares of natural forest in Peru. The project site includes national park reserves, and the investment will restore a 4000 hectare degraded buffer zone around these parks. The plan is to eventually produce "deforestation free" cocoa that will create jobs for local farmers and generate four million tonnes of certified carbon emission reductions. While financial returns are meant to be market rate, Althelia is backed by a USAID guarantee that halves the financial risk associated with the projects, a classic example of the collectivization of private sector risk.
4   How much does the materiality and the liveliness of living things and ecosystems explain the challenges economists and ecologists face in rendering biological diversity enterprising? This is a question that several people have asked me. Is the ultimate source of failure rooted in the unruliness of life on earth? There is no doubt in my mind that the unpredictability and uncertainties of living systems – human and nonhuman, always entangled – present

challenges for the hopes and dreams to enterprise nature. Not all "things" are as amenable to being staged as market, economic, or calculable objects – as one reviewer of this book nicely phrased. Through this book I identify moments when biological diversity scuppers instrumental reason, when, for example, financiers struggle to understand the financial risks of biodiversity loss. But biodiversity itself is also a material-semiotic object, forged out of the living, breathing array of lives on the planet and the scientific, political, cultural, and economic apparatuses of the West (see chapter 2). Any disruptions to the grand schemes of instrumental reason cannot be understood as achieved through the unruly nature of nature itself, outside of co-produced histories. Just as one would not want to understand socioecological relations as the result of inherent human characteristics like greed, it is dangerous to understand this process as the result of the natural characteristics of nonhuman natures. Further, as Callon (1998) argues, every market transaction is riddled with uncertainties, uncertainties parceled away in order to make exchange possible. Just because equivalence is difficult to achieve in relation to ecosystems (i.e. to decide that one ecosystem here can be compensated by an ecosystem over there), or that ecologists remain uncertain about which species are necessary for ecosystem functioning, does not mean that social agreement on these issues cannot be reached. As one academic legal scholar explained to me in an interview, establishing biodiversity-oriented trading schemes means that "you've got to be willing to accept some pretty ugly trade-offs" because it is clear that we are not "trading apples for apples." If we want to have market-based biodiversity policies like offsets, he went onto say, "we're basically going to accept imperfect proxies." The question is whether or not those proxies are deemed socially acceptable – a point nicely made by Morgan Robertson (2012).

5 Dipesh Chakrabarty (2007) uses the term "waiting room" to describe the historicist narrative of colonialism that suggested that non-Western peoples and nations were "not yet" ready to enter self-rule, they were "not yet" civilized enough. Chakrabarty argues that Western historicist narratives are teleological and universal – that they are based on an idea that we are all heading to essentially the same place, the same liberal democratic society, but some (Europeans, especially) get there before others do.

6 See http://www.un-ngls.org/IMG/pdf/Roundtable_4_Third_World_Network_ 25_September.pdf (last accessed March 13, 2016).

7 This is discussed in the preface of this book.

References

Adams, W.M. 2013. *Against extinction: the story of conservation.* Earthscan, London.

Adams, W.M. and K.H. Redford. 2010. Ecosystem services and conservation: a reply to Skroch and Lopez-Hoffman. *Conservation Biology* 24(1): 328–329.

Alatalo, R.V. 1981. Problems in the measurement of evenness in ecology. *Oikos* 37(2): 199–204.

Arkema, K.A., G. Guannel, G. Verutes, et al. 2013. Coastal habitats shield people and property from sea level rise and storms. *Nature Climate Change* 3(10): 913–918.

Armsworth, P.R., K.M.A. Chan, G.C. Daily, et al. 2007. Ecosystem-service science and the way forward for conservation. *Conservation Biology* 21(6): 1383–1384.

Arsel, M. and B. Büscher. 2012. Nature™ Inc: changes and continuities in neo-liberal conservation and market-based environmental policy. *Development and Change* 43(1): 53–78.

Bäckstrand, K. and E. Lövbrand. 2006. Planting trees to mitigate climate change: contested discourses of ecological modernization, green governmentality and civic environmentalism. *Global Environmental Politics* 6(1): 50–75.

Bagstad, K.J., D.J. Semmens, S. Waage, and R. Winthrop. 2013. A comparative assessment of decision-support tools for ecosystem services quantification and valuation. *Ecosystem Services* 5: 27–39.

Bakker, K. 2010. The limits of "neoliberal natures": debating green neoliber-alism. *Progress in Human Geography* 34(6): 715–735.

Balancing Biodiversity. 2008. Balancing biodiversity: towards a global incentive instrument for biodiversity preservation. Minutes of side event. Available at http://gdm.earthmind.net/files/Minutes-COP9-Balancing-Biodiversity-towards-a-global-incentive-instrument-for-biodiversity-preservation.pdf, last accessed September 19, 2011.

Enterprising Nature: Economics, Markets, and Finance in Global Biodiversity Politics,
First Edition. Jessica Dempsey.
© 2016 John Wiley & Sons, Ltd. Published 2016 by John Wiley & Sons, Ltd.

Baldwin, A. 2009. Carbon nullius and racial rule: race, nature and the cultural politics of forest carbon in Canada. *Antipode* 41(2): 231–255.

Barbier, E. B. and M. Rauscher. 1995. Policies to control tropical deforestation. In *Biodiversity loss*, C.A. Perrings, K.-G. Maler, C. Folke, C.S. Holling, and B.-O. Jansson (eds). Kluwer Academic Publishers, Dordrecht, The Netherlands, pp. 260–282.

Barbier, E., J.C. Burgess, and C. Folke. 1994. *Paradise lost? The ecological economics of biodiversity*. Earthscan, London.

Barnes, T.J. 1998. A history of regression: actors, networks, machines, and numbers. *Environment and Planning A* 30(2): 203–224.

Barnes, T.J. 2008. Making space for the economy: live performances, dead objects, and economic geography. *Geography Compass* 2(5): 1432–1448.

Barry, A. and D. Slater. 2002. Technology, politics and the market: an interview with Michel Callon. *Economy and Society* 31(2): 285–306.

Baskin, Y. 1994a. Ecologists dare to ask: how much does diversity matter? *Science* 264(5156): 202–203.

Baskin, Y. 1994b. Ecosystem function of biodiversity. *BioScience* 44(1): 657–660.

Baskin, Y. 1998. *The work of nature: how the diversity of life sustains us*. Island Press, Washington, DC.

Beck, U. 1992. *Risk society: towards a new modernity*. Sage, New Delhi.

Beirich, H. 2007. Federation for American Immigration Reform's hate filled track record. *Southern Poverty Law Centre*. Available at http://www.splcenter. org/get-informed/intelligence-report/browse-all-issues/2007/winter/the-teflon-nativists, last accessed February 22, 2015.

Berkes, F., C. Folke, and M. Gadgil. 1995. Traditional ecological knowledge, biodiversity, resilience and sustainability. In *Biodiversity conservation*, C.A. Perrings, K.-G. Maler, C. Folke, C.S. Holling, and B.-O. Jansson (eds). Kluwer Academic Publishers, Dordrecht, The Netherlands, pp. 269–287.

Berndt, C. and M. Boeckler. 2009. Geographies of circulation and exchange: constructions of markets. *Progress in Human Geography* 33(4): 535.

Berndt, C. and M. Boeckler. 2012. Geographies of marketization. In *The Wiley-Blackwell companion to economic geography*, T.J. Barnes, J. Peck, and E. Sheppard (eds). Wiley-Blackwell, Oxford.

Bernstein, S. 2002. *The Compromise of Liberal Environmentalism*. Columbia University Press, New York.

Biggs, R., K. Kotschy, A.M. Leitch, et al. 2012. Toward principles for enhancing the resilience of ecosystem services. *Annual Review of Environment and Resources* 37: 421–448.

Bio Assets. 2015. About Us. São Paulo, Brazil: Bio Assets. Available at http:// www.bioassets.com.br/quemsomos.php, last accessed February 2, 2015.

Blom, M., G. Bergsma, and M. Kortelan. 2008. *Economic instruments for biodiversity: setting up a biodiversity trading system in Europe*. Available at http://www.cedelft.eu/publicatie/economic_instruments_for_biodiversity/883, last accessed March 2, 2016.

Boyd, W., W.S. Prudham, and R.A. Schurman. 2001. Industrial dynamics and the problem of nature. *Society & Natural Resources* 14: 555–570.

Bracking, S. 2012. How do investors value environmental harm/care? Private equity funds, development finance institutions and the partial financialization of nature-based industries. *Development and Change* 43(1): 271–93.

Braun, B. 2002. *The intemperate rainforest: nature, culture, and power on Canada's West Coast.* University of Minnesota Press, Minneapolis.

Brenner, N., J. Peck, and N. Theodore. 2010a. After neoliberalization? *Globalizations* 7(3): 327–345.

Brenner, N., J. Peck, and N. Theodore. 2010b. Variegated neoliberalization: geographies, modalities, pathways. *Global Networks* 10(2): 182–222.

Brockerhoff, E.G., J. Hervé, J.A. Parrotta, et al. 2008. Plantation forests and biodiversity: oxymoron or opportunity? *Biodiversity and Conservation* 17(5) 925–951.

Brockington, D. 2002. *Fortress conservation: the preservation of the Mkomazi Game Reserve, Tanzania.* James Currey, Oxford.

Brockington, D. and R. Duffy. 2010. Capitalism and conservation: the production and reproduction of biodiversity conservation. *Antipode* 42(3): 469–84.

Brockington, D. and J. Igoe. 2006. Eviction for conservation: a global overview. *Conservation and Society* 4(3): 424–470.

Brockington, D., R. Duffy, and J. Igoe. 2008. *Nature unbound: conservation, capitalism and the future of protected areas.* Earthscan, Washington.

Brooks, T.M. 2011. Extinctions: consider all species. *Nature* 474(7351): 284.

Brown, G. and J. Roughgarden. 1995. An ecological economy: notes on harvest and growth. In *Biodiversity conservation*, C.A. Perrings, K.-G. Maler, C. Folke, C.S. Holling, and B.-O. Jansson (eds). Kluwer Academic Publishers, Dordrecht, The Netherlands, pp. 150–189.

Brown, K., D. Pearce, C. Perrings, and T. Swanson. 1992. *Economics and the conservation of global biological diversity.* World Bank and Global Environment Facility, New York.

Brown, W. 2005. *Edgework: critical essays on knowledge and politics.* Princeton University Press, Princeton.

Brown, W. 2015. *Undoing the demos: neoliberalism's stealth revolution.* MIT Press: Cambridge, MA.

Bullock, J.M., J. Aronson, A.C. Newton, et al. 2011. Restoration of ecosystem services and biodiversity: conflicts and opportunities. *Trends in Ecology & Evolution* 26(10): 541–549.

Bumpus, A.G. and D.M. Liverman. 2008. Accumulation by decarbonization and the governance of carbon offsets. *Economic Geography* 84(2): 127–155.

Burgess, J.C. 1995. The timber trade as cause of deforestation. In *Biodiversity conservation*, C.A. Perrings, K.-G. Maler, C. Folke, C.S. Holling, and B.-O. Jansson (eds). Kluwer Academic Publishers, Dordrecht, The Netherlands, pp. 226–244.

Burtis, P. 2008. Can bioprospecting save itself? At the vanguard of bioprospecting's second wave. *Journal of Sustainable Forestry* 25(3–4): 218–245.

Büscher, B. 2009. Letters of gold: enabling primitive accumulation through neoliberal conservation. *Human Geography* 2(3): 91–3.

Büscher, B., W. Dressler, and R. Fletcher. 2014. *Nature™ Inc: environmental conservation in the neoliberal age*. University of Arizona Press, Tucson.

Büscher, B., 2014. Nature on the Move I. In *Nature™ Inc: environmental conservation in the neoliberal age*, B. Büscher, W. Dressler and R. Fletcher (eds), pp. 183–204, University of Arizona Press, Tucson.

Büscher, B. and R. Fletcher. 2015. Accumulation by conservation. *New Political Economy* 20: 273–298.

Büscher, B., S. Sullivan, K. Neves, et al. 2012. Towards a synthesized critique of neoliberal biodiversity conservation. *Capitalism Nature Socialism* 23: 4–30.

Business Social Responsibility Network. 2008 *Measuring corporate impact on ecosystems: a comprehensive review of new tools*. Available at http://www.bsr.org/reports/BSR_EMI_Tools_Application1.pdf, last accessed March 2, 2016.

Business Social Responsibility Network. 2010. *Measuring environmental performance: the business case for new tools*. Available at http://www.bsr.org/en/our-insights/blog-view/measuring-environmental-performance-the-business-case-for-new-tools, last accessed March 2, 2016.

Business Social Responsibility Network. 2015. *Making the invisible visible: analytical tools for assessing business impacts and dependencies upon ecosystem services*. Available at http://www.bsr.org/reports/BSR_Analytical_Tools_for_Ecosystem_Services_2014.pdf, last accessed February 17, 2015.

Business Social Responsibility Network. 2013. About BSR. Available at www.bsr.org, last accessed May 2, 2013.

Butchart, S.H.M., A.J. Stattersfield, L.A. Bennun, et al. 2004. Measuring global trends in the status of biodiversity: red list indices for birds. *PLoS Biology* 2(12): e383.

Butchart, S.H.M., M. Walpole, B. Collen, et al. 2010. Global biodiversity: indicators of recent declines. *Science* 328(5982): 1164–1168.

Caldecott, B. 2011. Protecting and restoring biodiversity will require private capital. *Guardian*, February 15, 2011.

Callon, M. 1998. An essay on framing and overflowing: economic externalities revisited by sociology. In *The laws of the market*, M. Callon (ed). Blackwell Publishers, Oxford, pp. 244–269.

Callon, M. and F. Muniesa. 2005. Peripheral vision economic markets as calculative collective devices. *Organizational Studies* 26(8): 1229–1250.

Campanale, M. 2008. Private capital markets and forests: key questions in marketing forest carbon investments to institutions. Presentation at the "Biodiversity and Ecosystem Finance" conference, London, November 2008.

Campbell, E. 2006. The case of the $150,000 fly. Mitigation: the Ecosystem Marketplace's daily coverage of the 2006 mitigation/conservation banking conference. Washington, DC: Ecosystem Marketplace. Available at http://moderncms.ecosystemmarketplace.com/repository/moderncms_documents/mitigation_news_4.26.06.1.1.pdf, last accessed September 16, 2014.

Carpenter, S.R., R. DeFries, T. Dietz, et al. 2006. Millennium Ecosystem Assessment: research needs. *Science* 314(5797): 257.

Carroll, N., J. Fox, and R. Bayon. 2008. *Conservation and biodiversity banking: a guide to setting up and running biodiversity credit trading systems.* Earthscan, London.

Carson, R. 1962. *Silent Spring.* Houghton Mifflin, Boston.

Castree, N. 2003a. Bioprospecting: from theory to practice (and back again). *Transactions of the Institute of British Geographers* 28(1): 35–55.

Castree, N. 2003b. Commodifying what nature? *Progress in Human Geography* 27(3): 273–297.

Castree, N. 2008a. Neoliberalising nature: the logics of deregulation and reregulation. *Environment and Planning A* 40(1): 131–152.

Castree, N. 2008b. Neoliberalising nature: processes, effects, and evaluations. *Environment and Planning A* 40(1): 153–173.

Castree, N. and B. Braun. 2001. *Social nature: theory, practice, and politics.* Blackwell Publishers Oxford, Malden, MA.

Cavanagh, C. and T.A. Benjaminsen. 2014. Virtual nature, violent accumulation: The "spectacular failure" of carbon offsetting at a Ugandan national park. *Geoforum* 56: 55–65.

CBD Alliance. 2010. *Top 10 for COP 10.* Available from author.

Ceballos, G. P.R Ehrlich, A.D. Barnosky, A. García, R.M. Pringle, and T.M. Palmer. 2015. Accelerated modern human-induced species losses: Entering the sixth mass extinction. *Science Advances* 1, e1400253.

Chakrabarty, D. 2007. *Provincializing Europe: postcolonial thought and historical difference (2nd edition).* Princeton University Press, Princeton.

Chaloupka, W. 2003. There must be some way out of here: strategy, ethics, and environmental politics. In *A political space: reading the global through Clayoquot Sound,* W. Magnusson and K. Shaw (eds), pp. 67–90.

Chan, K. 2004. Conservation finance: the design of conservation investments. *Centre for Conservation Biology Newsletter* 16(1), fall 2004. Available at http://www.stanford.edu/group/CCB/Pubs/Updatepdfs/Update%20fall04. pdf, last accessed September 15, 2011.

Chapin, M. 2004. A challenge to conservationists. *WorldWatch Magazine* (November/December edition).

Chapin, F.S., E.D. Schultze, and H.A. Mooney. 1992. Biodiversity and ecosystem processes. *Trends in Ecology and Evolution* 7(4): 107–108.

Chatty, D. and M. Colchester (eds). 2002. *Conservation and mobile indigenous peoples: displacement, forced settlement and sustainable development.* James Berghan, Oxford.

Chiarucci, A., G. Bacaro, and S.M. Scheiner. 2011. Old and new challenges in using species diversity for assessing biodiversity. *Philosophical Transactions of the Royal Society of London. Series B, Biological Sciences* 366(1576): 2426–2437.

Choucroun, K. 2010. Biodiversity and profit can go hand in hand. *Guardian,* October 6, 2010. Available at http://www.guardian.co.uk/environment/2010/oct/06/biodiversity-business-opportunity, last accessed October 7, 2010.

Ciriacy-Wantrup, S.V. 1952. *Resource conservation: economics and policies.* University of California Press, Berkeley.

Clapp, R.A. and C. Crook. 2002. Drowning in the magic well: Shaman Pharmaceuticals and the elusive value of traditional knowledge. *The Journal of Environment and Development* 11(1): 79–102.

Clare, S., N. Krogman, L. Foote, and N. Lemphers. 2011. Where is the avoidance in the implementation of wetland law and policy? *Wetlands Ecology Management* 19(2): 165–182.

Clayton, D. 2000. Governmentality. In *Dictionary of human geography (4th edition)*, R.J. Johnston, D. Gregory, G. Pratt, and M. Watts (eds). Blackwell, London, p. 318.

Cleary, D. 2006. The questionable effectiveness of science spending by international conservation oganizations in the tropics. *Conservation Biology* 20(3): 733–738.

Coase, R. H. 1960. The problem of social cost. *Journal of Law and Economics* 3(1): 1–44.

Code, L. 2006. *Ecological thinking: the politics of epistemic location.* Oxford University Press, Oxford.

Collard, R.-C. and J. Dempsey. 2013. Life for sale? The politics of lively commodities. *Environment and Planning A* 45: 2682–2699.

Collard, R.-C., J. Dempsey, and J. Sundberg. 2015. Disentangling the multiple and contradictory logics of Nature™ Inc. *Environment and Planning A* 47: 2389–2408.

Collard, R.-C., J. Sundberg, and J. Dempsey. 2014. Abundant futures manifesto. *Annals of the Association of American Geographers* 105(2): 322–330.

Collen, B., J. Loh, S. Whitmee, et al. 2009. Monitoring change in vertebrate abundance: the living planet index. *Conservation Biology* 23(2): 317–327.

Confino, J. 2011. The banker trying to put a value on nature. *Guardian,* May 16, 2011. Available at http://www.guardian.co.uk/sustainable-business/pavan-sukhdev-valuing-biodiversity-ecosystem-services, last accessed September 15, 2011.

Connelly, M.J. 2008. *Fatal misconception: the struggle to control world population.* Harvard University Press, Cambridge, Boston.

Convention on Biological Diversity. 2009. *Draft global biodiversity outlook 3.* Available from author.

Convention on Biological Diversity. 2010a. *Global biodiversity outlook (GBO).* Available at http://gbo3.cbd.int/resources.aspx, last accessed September 16, 2011.

Convention on Biological Diversity. 2010b. *Jakarta charter on business and biodiversity.* Available at http://www.cbd.int/doc/business/jakarta-charter-busissness-en.pdf, last accessed September 15, 2011.

Convention on Biological Diversity 2010c. Policy options concerning innovative financial mechanisms. Available at http://www.cbd.int/doc/meetings/wgri/wgri-03/official/wgri-03-08-en.doc, last accessed May 7, 2013.

Cooper, D. 1991. Overcoming the obstacles to a global agreement on conservation and sustainable use of biodiversity. In *Biodiversity: social and ecological perspectives*, Shiva et al. (eds). Zed Books, London and World Rainforest Movement, Montevideo, Uruguay.

Cooper, M. 2008. *Life as surplus*. University of Washington Press, Seattle.

Corbera, E. and K. Brown. 2010. Offsetting benefits? Analyzing access to forest carbon. *Environment and Planning A* 42(7): 1739–1761.

Corbera, E., N. Kosoy, and M. Martínez-Tuna. 2007. Equity implications of marketing ecosystem services in protected areas and rural communities: case studies from Meso-America. *Global Environmental Change* 17(3): 365–380.

Corporate Ecosystem Services Review. 2008. *The corporate ecosystem services review: guidelines for identifying business risks and opportunities arising from ecosystem change*. World Resources Institute, Washington, DC.

Costanza, R. 2006. Nature: ecosystems without commodifying them. *Nature* 443(7113): 749.

Costanza, R. and H. Daly. 1987. Toward an ecological economics. *Ecological Modelling* 38(1–2): 1–7.

Costanza, R., M. Kemp, and W. Boynton. 1995. Scale and biodiversity in coastal and estuarine ecosystems. In *Biodiversity conservation*, C.A. Perrings, K.-G. Maler, C. Folke, et al. (eds). Kluwer Academic Publishers, Dordrecht, The Netherlands, pp. 84–125.

Costanza, R., R. D'Arge, R. de Groot, et al. 1997. The value of the world's ecosystem services and natural capital. *Nature* 387(6630): 253–260.

Costanza, R, R. d'Arge, R. de Groot, et al. 1998. The value of ecosystem services: putting the issues in perspective. *Ecological Economics* 25: 67–72.

Creagh, S. 2009. Forest-CO2 scheme will draw organized crime: Interpol. May 29, 2009. Nusa Dua, Indonesia: Reuters. Available at http://www.reuters.com/article/2009/05/29/us-indonesia-carbon-crime-sb-idUSTRE54S1DS20090529, last accessed February 2, 2015.

Credit Suisse, World Wildlife Fund, and McKinsey & Company. 2014. *Conservation finance: moving beyond donor funding toward an investor-driven approach*. Available at https://www.credit-suisse.com/media/cc/docs/responsibility/conservation-finance-en.pdf (last accessed 23 November, 2015).

Cronon, W. 1992. *Nature's Metropolis*. W.W. Norton & Co., New York.

Cronon, W. 1995. The trouble with wilderness. In *Uncommon ground: rethinking the human place in nature*, W. Cronon (ed.). W.W. Norton & Co., New York, pp. 69–90.

Cusens, J., S.D. Wright, P.D. McBride, and L.N. Gillman. 2012. What is the form of the productivity-animal-species-richness relationship? A critical review and meta-analysis. *Ecology* 93(10): 2241–2252.

Daily, G.C. 1997. *Nature's services: societal dependence on natural ecosystems*. Island Press, Washington, DC.

Daily, G.C. and K. Ellison. 2002. *The new economy of nature: the quest to make conservation profitable*. Island Press, Washington, DC.

Dalton, R. 2004. Bioprospects less than golden. *Nature* 429(6992): 598–600.

Daly, H.E. (ed.). 1973. *Toward a steady-state economy*. W. H. Freeman, San Francisco.

Daly, H.E. 1977. *Steady-state economics: the economics of equilibrium and moral growth*. WH Freeman: San Francisco.

Daly, H.E. 1991. *Steady state economics (2nd edition)*. Island Press, Washington, DC.

Dasgupta, P. 1995. Foreword. In *Biodiversity loss*, C.A. Perrings, K.-G. Maler, C. Folke, C.S. Holling, and B.-O. Jansson (eds). Kluwer Academic Publishers, Dordrecht, The Netherlands, pp. vii–x.

Dempsey, J. 2010. Tracking grizzly bears in British Columbia's environmental politics. *Environment and Planning A* 42(5): 1138–1156.

Dempsey, J. 2009. The 2010 target will not be met! *Third World Resurgence* 231–232: 29–32. Available at http://www.twn.my/title2/resurgence/2009/231-232/cover4.htm, last accessed February 29, 2016.

Dempsey, J. and M. Robertson. 2012. Ecosystem services: tensions, impurities, and points of engagement within neoliberalism. *Progress in Human Geography* 36: 758–779.

Dempsey, J. and D.C. Suarez. 2016. Arrested development? The promises and paradoxes of "selling nature to save it." *Annals of the Association of American Geographers* 106(3): 653–671.

Dirzo, R., H.S. Young, M. Galetti, G. Ceballos, et al. 2014. Defaunation in the Anthropocene. *Science* 345: 401–406.

Dixon, J., L. Scura, and T. Hof. 1995. Valuation of a marine resource. In *Biodiversity conservation*, C.A. Perrings, K.-G. Maler, C. Folke, C.S. Holling, and B.-O. Jansson (eds). Kluwer Academic Publishers, Dordrecht, The Netherlands, pp. 120–137.

Dornelas, M., M.A. Kosnik, B. McGill, et al. 2013. Quantifying temporal change in biodiversity: challenges and opportunities. *Proceedings. Biological Sciences/ the Royal Society* 280(1750): 20121931.

Dorsey, M.K. 2007. Carbon trading won't work. April 1, 2007. *LA Times*. Available at http://www.latimes.com/la-op-dorsey1apr01-story.html, last accessed February 2, 2015.

Dulvy, N. K. and S. L. Fowler. 2009. Global threat status of the world's sharks, rays and chimaeras. *Bleeding the oceans dry: the overfishing and decline of global sharks stocks*. WildAid, London: 6–7.

Dulvy, N.K., S.L. Fowler, J.A. Musick, et al. 2014. Extinction risk and conservation of the world's sharks and rays. *eLife 3*.

Dunn, E.C. 2007. Of pufferfish and ethnography: plumbing new depths in economic geography. In *Politics and practice in economic geography*, A. Tickell, E. Sheppard, J. Peck, and T.J. Barnes (eds). Sage, London, pp. 82–92.

Earthmind. 2013. About us. Available at http://earthmind.net/aboutus/, last accessed May 3, 2013.

ECO. 2005. Biodiversity provides public benefits, it is a public asset, not a service. *ECO* 14(3): 1. Available from author.

Economist. 2005. Are you being served? *Economist*, April 1, 2005. Available at http://www.economist.com/node/3886849, last accessed February 27, 2016.

Economist. 2008. Where the wild things are. *Economist*, October 2, 2008. Available at http://www.economist.com/node/12332923, last accessed September 15, 2011.

Ecosystem Marketplace. 2013. Overview: about the Ecosystem Marketplace. Available at www.ecosystemmarketplace.com, last accessed May 10, 2013.

Egziabher, T. 2002. The human individual and community in the conservation and sustainable use of biological resources. Darwin Lecture at the Annual Darwin Initiative meeting, London. http://www.nyeleni.org/IMG/pdf/Tewolde_Darwin_Lecture2002.pdf, last accessed November 1, 2013.

Ehrlich, P. 1968. *The population bomb*. Ballantine Books, New York.

Ehrlich, P. 1988. The loss of diversity: causes and consequences. In *Biodiversity*, E.O. Wilson (ed.). National Academy Press, Washington, DC, pp. 21–26.

Ehrlich, P. 1997. Foreword. In *The work of nature: how the diversity of life sustains us*, Y. Baskin (author). Island Press, Washington, pp. ix–xii.

Ehrlich, P. 1998. Foreword: Of keystone complexes and nature's services. In Y. Baskin, *The work of nature: how the diversity of life sustains us*. Island Press, Washington, DC, pp. ix–xii.

Ehrlich, P.R. and A. Ehrlich. 1970. *Population, resources, environment: issues in human ecology*. W.H. Freeman, San Francisco.

Ehrlich, P. and A. Ehrlich. 1981. *Extinction: the causes and consequences of the disappearance of species*. Ballantine Books, New York.

Ehrlich, P. and H.A. Mooney. 1983. Extinction, substitution and ecosystem services. *BioScience* 33(4): 248–254.

Ehrlich, P. and H.A. Mooney. 1997. Ecosystem services: a fragmentary history. In *Ecosystem services*, G.C. Daily (ed.). Island Press, Washington, DC, pp. 11–19.

EIRIS. 2010. 'COP' out? Biodiversity loss and the risk to investors. Available at http://www.eiris.org/files/research%20publications/Biodiversity2010.pdf, last accessed September 11, 2011.

Elton, C.S. 1958. *The ecology of invasions by plants and animals*. John Wiley and Sons, New York.

Emel, J. and M.T. Huber. 2008. A risky business: mining, rent and the neoliberalization of "risk." *Geoforum* 39(3): 1393–1407.

Environmental Leader. 2012. Forest footprint disclosure project to merge with CDP. June 14, 2012. Available at http://www.environmentalleader.com/2012/06/14/forest-footprint-disclosure-project-to-merge-with-cdp/, last accessed October 3, 2014.

Equator Principles. 2013. About the Equator Principles. Available at: http://www.equator-principles.com/index.php/about-ep/about-ep, last accessed May 3, 2013.

Escobar, A. 1998. Whose knowledge, whose nature? Biodiversity, conservation, and the political ecology of social movements. *Journal of Political Ecology* 5(1): 53–82.

ETC Group. 2013. Biopiracy. Available at http://www.etcgroup.org/issues/patents-biopiracy, last accessed April 30, 2013.

Evans, M., H. Possingham, and K. Wilson. Extinctions: conserve not collate. *Nature* 474 (7351): 284.

Ewald, F. 1991. Insurance and risk. In *The Foucault effect: studies in governmentality*, G. Burchell, C. Gordon, and P. Miller (eds). University of Chicago Press, Chicago, pp. 197–210.

Fairhead, J. and M. Leach. 1995. False forest history, complicit social analysis: Rethinking some West African environmental narratives. *World Development* 23(6): 1023–1035.

Farnham, T.J. 2007. *Saving nature's legacy: origins of the idea of biological diversity.* Yale University Press, New Haven.

Felli, R. 2014. On climate rent. *Historical Materialism* 22(3–4): 251–280.

Ferguson, J. 1994. *The anti-politics machine: "development," depoliticization, and bureaucratic power in Lesotho.* Cambridge University Press, Cambridge.

Ferguson, J. 2005. Seeing like an oil company: space, security, and global capital in neoliberal Africa. *American Anthropologist* 107(3): 377–382.

Firn, R.D. 2003. Bioprospecting — why is it so unrewarding? *Biodiversity and Conservation* 12: 207–216.

Fletcher, R. 2010. Neoliberal environmentality: towards a poststructuralist political ecology of the conservation debate. *Conservation and Society* 8(3): 171.

Fletcher, R., J. Breitling. 2012. Market mechanism or subsidy in disguise? Governing payment for environmental services in Costa Rica. Geoforum, *The Global Rise and Local Implications of Market-Oriented Conservation Governance* 43: 402–411.

Fletcher, R., W. Dressler, and B. Büscher, B. 2014. Nature Inc: the new frontiers of environmental conservation. In *Nature™ Inc: environmental conservation in the neoliberal age*, B. Büscher, W. Dressler, and R. Fletcher (eds). University of Arizona Press, pp. 3–21.

FAO 2010. Second report on the state of the world's plant genetic resources for food and agriculture. FAO: Rome, Italy.

Forsyth, T. 2002. *Critical political ecology.* Routledge, London.

Foster, J. 1993. Economics and the self-organisation approach: Alfred Marshall revisited? *The Economic Journal* 103(419): 975–991.

Foucault, M. 1977. The confession of the flesh: 1977 interview. In *Power/knowledge: selected interviews and other writings*, C. Gordon (ed.), 1980. Harvester, London, pp. 194–228.

Foucault, M. 1991. Governmentality. In *The Foucault effect*. G. Burchell, C. Gordon, and P. Miller (eds). University of Chicago Press, Chicago.

Foucault, M. 1995 [1975]. *Discipline and punish: the birth of the prison.* Random House, New York.

Foucault, M. 2003. *Society must be defended: lectures at the Collège de France, 1975–1976.* David Macey (trans). Picador, New York.

Foucault, M. 2008. *The birth of biopolitics: lectures at the Collège de France, 1978–79.* Palgrave Macmillan, New York.

Fraser, N. 2014. Behind Marx's hidden abode. *New Left Review* 86: 55–72.

Fukuyama, F. 1992. *The end of history and the last man.* New York, Free Press.

Georgescu-Roegen, N. 1971. *The entropy law and the economic process.* Harvard University Press, Cambridge, Massachusetts.

Gibson-Graham, J.K. 1996. *The end of capitalism (as we knew it).* Blackwell Publishers, Cambridge and Oxford.

Gilbertson, T. and O. Reyes. 2009. *Carbon trading – how it works and why it fails.* Dag Hammarskjöld Foundation. Available at https://www.tni.org/en/publication/carbon-trading-how-it-works-and-why-it-fails, last accessed March 2, 2016.

Global Canopy Programme. (GCP). 2010a. *The little book of biodiversity finance*. Global Canopy Programme, London.

Global Canopy Programme (GCP), IPAM, Fauna & Flora International (FFI) and UNEP Finance Initiative (UNEP FI). 2014. Stimulating interim demand for REDD+ emission reductions: the need for a strategic intervention from 2015 to 2020, (GCP), Oxford, UK; the Amazon Environmental Research Institute, Brasília, Brazil; (FFI), Cambridge, UK; and UNEP FI, Geneva, Switzerland.

Goldman, M. 2005. *Imperial nature: the World Bank and struggles for social justice in the age of globalization.* Yale University Press, New Haven.

Gómez-Baggethun, E., R. De Groot, P.L. Lomas, and C. Montes. 2010. The history of ecosystem services in economic theory and practice: from early notions to markets and payment schemes. *Ecological Economics* 69(6): 1209–1218.

Gorelick, R. 2011. Commentary: Do we have a consistent terminology for species diversity? The fallacy of true diversity. *Oecologia* 167(4): 885–888.

Gotelli, N.J. and R.K. Colwell. 2001. Quantifying biodiversity: procedures and pitfalls in the measurement and comparison of species richness. *Ecology Letters* 4(4): 379–391.

Green Development Mechanism. 2010a. *Making the case for a Green Development Mechanism (GDM)*. Presentation to the UNEP Governing Council, Bali, February 2010. Available at http://gdm.earthmind.net/files/gdm-bali-unep-side-event-presentation.pdf, last accessed September 15, 2011.

Green Development Mechanism. 2010b. *Lobby document circulated at WGRI 3*. Available from author.

Green Power Conferences. 2013a. About us: Green Power Conferences is the market leader in renewable energy conferences. Available at http://www.greenpowerconferences.com/home/aboutus, last accessed May 3, 2013.

Green Power Conferences. 2013b. Webinars. Available at http://www.greenpowerconferences.com/home/webinars, last accessed May 3, 2013.

Gregory, D. 2001. Postcolonialism and the production of nature. In *Social nature: theory, practice, and politics.* N. Castree and B. Braun (eds). Blackwell, Oxford. pp. 84–98.

Griffiths, T. 2008. *Seeing "REDD"? Forests, climate change mitigation and the rights of indigenous peoples and local communities.* Forest Peoples Programme. Available at http://www.forestpeoples.org/sites/fpp/files/publication/2010/08/seeingreddupdatedraft3dec08eng.pdf, last accessed September 15, 2011.

Griggs, A., Z. Cullen, J. Foxall, and R. Strumpf. 2009. *Linking shareholder and natural value. Managing biodiversity and ecosystem service risk in companies with an agricultural supply chain.* Fauna & Flora International, United Nations Environment Program Finance Initiative, and Fundacao Getullo Vargas.

Guerry, A.D., M.H. Ruckelshaus, K. Arkema, et al. 2012. Modelling benefits from nature: using ecosystem services to inform coastal and marine spatial planning. *International Journal of Biodiversity Science, Ecosystem Services and Management* 8:107–121.

Guha, R. 2006. Radical American environmentalism and wilderness preservation: a third world critique. In *Moral issues in global perspective*, C.M. Koggel (ed.). Broadview Press, Peterborough, pp. 253–262.

Guyer, J. and P. Richards. 1996. The invention of biodiversity: social perspectives on the management of biological variety in Africa. *Africa: Journal of the International African Institute* 66(1): 1–13.

Hacking, I. 1990. *The taming of chance*. Cambridge University Press.

Hannah, M.G. 2011. Biopower, life and left politics. *Antipode* 43(4): 1034–1055.

Haraway, D. 1988. Situated knowledges. *Feminist Studies* 14(3): 575–599.

Haraway, D.J. 1991. A cyborg manifesto: science, technology and socialist-feminism in the late twentieth century, in *Simians, cyborgs and women: the reinvention of nature*. Routledge, New York, pp. 149–181.

Haraway, D.J. 1992. The promises of monsters: a regenerative politics for inappropriate/d others. In L. Grossberg, C.Nelson, and P.A. Treichler (eds), *Cultural Studies*. Routledge, New York, pp. 295–337.

Haraway, D.J. 1994. A game of cat's cradle: science studies, feminist theory, cultural studies. *Configurations* 2.1: 59–71.

Haraway, D.J. 1997. *Modest_Witness@Second_Millennium. FemaleMan_Meets_OncoMouse: feminism and technoscience*. Routledge, London.

Haraway, D.J. 2003. *The companion species manifesto: Dogs, people, and significant otherness*. Prickly Paradigm Press, Chicago.

Haraway, D.J. 2008. *When species meet*. University of Minnesota Press, Minneapolis.

Hari, J. 2010. The wrong side of green. *The Nation*, March 22, 2010. Available at http://www.thenation.com/article/wrong-kind-green, last accessed September 16, 2011.

Harris, J. 2009. Copenhagen climate conference: the key players. *Guardian*, November 30, 2009. Available at http://www.guardian.co.uk/environment/2009/nov/30/copenhagen-key-players, last accessed September 15, 2011.

Harvey, D. [1974] 2001. Population, resources and the ideology of science. In *Spaces of capital: towards a critical geography*. Routledge, London, pp. 38–67.

Harvey, D. 1996. *Justice, nature and the geography of difference*. Blackwell.

Harvey, D. 2001. *Spaces of capital: towards a critical geography*. Taylor & Francis, New York.

Harvey, D. 2005. *A brief history of neoliberalism*. Oxford University Press, New York.

Harvey, F. 2010. Saving species: bad for biodiversity is often bad for business. *Financial Times*, October 1, 2010. Available at http://www.ft.com/intl/cms/s/0/ce6b8e02-cceb-11df-9bf0-00144feab49a.html#axzz1YM3jF04H, last accessed September 16, 2011

Hayden, C. 2003. *When nature goes public: the making and unmaking of bioprospecting in Mexico*. Princeton University Press, Princeton.

He, F. and S.P. Hubbell. 2011. Species-area relationships always overestimate extinction rates from habitat loss. *Nature* 473(7347): 368–371.

Heynen, N., J. McCarthy, S. Prudham, and P. Robbins. 2007. *Neoliberal environments*. Routledge, New York.

Hobbs, R.J., E. Higgs, and J.A. Harris. 2009. Novel ecosystems: implications for conservation and restoration. *Trends in Ecology & Evolution* 24(11): 599–605.

Hoffman, M., C. Hilton-Taylor, A. Angulo, et al. 2010. The impact of conservation on the world's vertebrates. *Science* 330(6010): 1503–1509.

Holling, C.S. 1973. Resilience and stability of ecological systems. *Annual Reviews of Ecology and Systematics* (4):1–23.

Holling, C.S. 1987. Simplifying the complex: the paradigms of ecological function and structure. *European Journal of Operational Research* 30(2): 139–146.

Holling, C.S., D.W. Schindler, B.W. Walker, and J. Roughgarden. 1995. Biodiversity in the functioning of ecosystems: an ecological synthesis. In *Biodiversity conservation*, C.A. Perrings, K.-G. Maler, C. Folke, et al. (eds). Kluwer Academic Publishers, Dordrecht, The Netherlands, pp. 44–83.

Holling, C.S., F. Berkes, and C. Folke. 1998. Science, sustainability and resource management. In Berkes, F. and C. Folke (eds). *Linking social and ecological systems: Management practices and social mechanisms for building resilience*. Cambridge University Press, Cambridge, pp. 342–362.

Horkheimer, M. and T. Adorno. 1944. The concept of enlightenment. *Dialectic of enlightenment*. Stanford University Press, Palo Alto.

Hough, P. and M. Robertson. 2009. Mitigation under Section 404 of the Clean Water Act: where it comes from, what it means. *Wetlands Ecology and Management* 17(1): 15–33.

Hutchinson, G.E. 1959. Homage to Santa Rosalia; or, why are there so many kinds of animals? *American Naturalist* 93(May–June): 145–159.

Huwyler, F. and J. Tobin. 2014. *Conservation finance: moving beyond donor funding toward an investor-driven approach*. January 2014. Credit Suisse, WWF, and McKinsey & Company. Available at http://www.longfinance.net/programmes/london-accord/la-reports.html?view=report&id=422, last accessed February 3, 2015.

IBAT. 2011. Homepage. Available at https://www.ibatforbusiness.org/, last accessed September 19, 2011.

Igoe, J. 2012. Nature on the move II: contemplation becomes speculation. *New Proposals: Journal of Marxism and Interdisciplinary Inquiry* 6: 37–49.

Igoe, J. and D. Brockington. 2007. Neoliberal conservation: a brief introduction. *Conservation and Society* 5(4): 432.

Igoe, J., K. Neves, and D. Brockington. 2010. A spectacular eco-tour around the historic bloc: theorising the convergence of biodiversity conservation and capitalist expansion. *Antipode* 42(3): 486–512.

IISD. 2010. *Summary of the third meeting of the ad hoc open-ended working group on review of implementation of the convention on biological diversity*. Available at http://www.iisd.ca/vol09/enb09519e.html, last accessed September 15, 2010.

International Finance Corporation. 2010. *Forest-backed bonds proof of study concept*. Issue Brief. Washington, DC: World Bank Group. Available at http://www.ifc.org/

wps/wcm/connect/4957218048855503b524f76a6515bb18/IFC_Breif_Forest_
web.pdf?MOD=AJPERES&CACHEID=4957218048855503b524f76a6515bb18,
last accessed February 2, 2015.

Isbell, F., J. van Ruijven, A. Weigelt, et al. 2011. High plant diversity is needed to
maintain ecosystem services. *Nature* 477(7363): 199.

ISIS Asset Management. 2004. Is biodiversity a material risk for companies? An
assessment of the exposure of FTSE sectors to biodiversity risks. Available at https://
www.globalnature.org/bausteine.net/f/6645/FC20Biodiversity20Report20FINAL.
pdf?fd=2, last accessed March 2, 2016.

IUCN. 1980. *World conservation strategy: living resources conservation for sustainable development.* IUCN, UNEP and WWF.

IUCN. 2011. Press release. Available at http://www.iucn.org/about/union/com
missions/ceesp/ceesp_publications/?7257/The-Wetland-Carbon-Partnership-
continues-for-a-3rd-Year, last accessed September 19, 2011.

Jackson, C. 1995. Radical environmental myths: a gender perspective. *New Left
Review* 201: 124–140.

Jacobs, R. 2013. The forest mafia: how scammers steal millions through carbon
markets. *Atlantic.* October 11, 2013. Available at http://www.theatlantic.
com/international/archive/2013/10/the-forest-mafia-how-scammers-steal-
millions-through-carbon-markets/280419/, last accessed February 3, 2014.

Jacquet, J., J. Hocevar, S. Lai, et al. 2010. Conserving wild fish in a sea of market-
based efforts. *Oryx* 44(1): 45–56.

Jessop, B. 2007. From micro-powers to governmentality: Foucault's work on
statehood, state formation, statecraft and state power. *Political Geography*
26(1): 34–40.

Johnson, L. 2013. Catastrophe bonds and longevity risk: entanglements of
capital and biopolitical rule. *Geoforum* 45: 30–40.

Johnson, N. 2014. Why Vandana Shiva is so right and yet so wrong. *Grist,*
August 20, 2014. Available at http://grist.org/food/vandana-shiva-so-right-
and-yet-so-wrong/, last accessed February 27, 2016.

Johnson, K.A., S. Polasky, E. Nelson, and D. Pennington. 2012. Uncertainty in
ecosystem services valuation and implications for assessing land use tradeoffs:
an agricultural case study in the Minnesota River Basin. *Ecological Economics*
79: 71–79.

Kareiva, P., H. Tallis, T. Ricketts, G. et al. 2011. *Natural capital: theory and
practice of mapping ecosystem services.* Oxford University Press, Oxford.

Kareiva, P., M. Marvier, and R. Lalasz. 2012. Conservation in the Anthropocene.
Breakthrough Journal (Winter 2012). Available at http://thebreakthrough.
org/index.php/journal/past-issues/issue-2/conservation-in-the-anthropocene/,
last accessed February 22, 2015.

Kenny, A. 2009. Will US stimulus lift mitigation banks? *Ecosystem Marketplace,*
February 1, 2009. Available at http://www.ecosystemmarketplace.com/pages/
dynamic/article.page.php?page_id=6510&section=home&eod=1, last accessed
August 31, 2009.

Kettlewell, C., V. Bouchard, D. Porej, et al. 2008. An assessment of wetland
impacts and compensatory mitigation in the Cuyahoga River Watershed,
Ohio, USA. *Wetlands* 28(1): 57–67.

Klein, N. 2010. Bolivia's fight for survival can help save democracy too. *Guardian,* April 22, 2010. Available at http://www.guardian.co.uk/comment isfree/cifamerica/2010/apr/22/how-bolivia-transformation-could-change-world, last accessed February 22, 2015.

Klein, N. 2015. *This changes everything: capitalism vs. the climate.* Simon & Schuster, New York.

KPMG. 2011. Sustainable insight: the nature of ecosystem service risks for business. Available at http://www.unepfi.org/fileadmin/documents/Sustainable_ Insight_May_2011.pdf, last accessed February 17, 2015.

Kolbert, E. 2014. The sixth extinction: an unnatural history. Henry Holt and Company. New York.

Kosoy, N. and E. Corbera. 2010. Payments for ecosystem services as commodity fetishism. *Ecological Economics,* 69: 1228–1236.

Kosoy, N., M. Martinez-Tuna, R. Muradian, and J. Martinez-Alier. 2007. Payments for environmental services in watersheds: insights from a comparative study of three cases in Central America. *Ecological Economics* 61(2): 446–455.

Kremen, C. 2005. Managing ecosystem services: what do we need to know about their ecology? *Ecology Letters* 8(5): 468–479.

Kremen, C. 2015. Reframing the land-sparing/land-sharing debate for biodiversity conservation. *Annals of the New York Academy of Sciences* 1355: 52–76.

Kremen, C. and R.S. Ostfeld. 2005. A call to ecologists: measuring, analyzing, and managing ecosystem services. *Frontiers in Ecology* 3(10): 540–548.

Kroeger, T. and F. Casey. 2007. An assessment of market-based approaches to providing ecosystem services on agricultural lands. *Ecological Economics* 64(2): 321–332.

Kyte, R. 2012. Rio's buzzing about natural capital accounting. *Voices.* World Bank. Available at http://blogs.worldbank.org/voices/rios-buzzing-about-natural-capital-accounting, last accessed May 7, 2013.

Lansing, D.M. 2011. Realizing carbon's value: discourse and calculation in the production of carbon forestry offsets in Costa Rica. *Antipode* 43: 731–753.

Larner, W. 2000. Neo-liberalism: policy, ideology, governmentality. *Studies in Political Economy* 63: 5–25.

Larner, W. 2007. Neoliberal governmentalities. In Heynen, N., J. McCarthy, S. Prudham, and P. Robbins (eds). *Neoliberal environments.* Routledge, New York, pp. 217–220.

Latour, B. 1987. *Science in action.* Harvard University Press, Boston.

Latour, B. 2004. *Politics of Nature.* Harvard University Press, Boston.

Leahy, S. 2010. Market-based conservation brewing in Nairobi. *TierraAmérica, Inter Press Service,* May 31, 2010. Available at http://www.ipsnews.net/2010/ 06/environment-market-based-conservation-brewing-in-nairobi/, last accessed March 2, 2016.

Legg, S. 2011. Assemblage/apparatus: using Deleuze and Foucault. *Area* 43(2): 128–133.

Leinster, T. and C.A. Cobbold. 2012. Measuring diversity: the importance of species similarity. *Ecology* 93(3): 477–489.

Lemke, T. 2001. "The birth of bio-politics": Michel Foucault's lecture at the Collège de France on neo-liberal governmentality. *Economy and Society* 30(2): 190–207.

Leopold, A. [1949] 1966. *A sand county almanac.* Oxford University Press, New York.

Lépinay, V.A. 2007. Decoding finance: articulation and liquidity around a trading room. In *Do economists make markets? On the performativity of economics,* MacKenzie, D.A., F. Muniesa, and L. Siu (eds). Princeton University Press, Princeton, pp. 87–127.

Levitt, T. 2010. Pavan Sukhdev: you can have progress without GDP-led growth. *Ecologist,* January 22, 2010. Available at http://www.theecologist.org/ Interviews/402389/pavan_sukhdev_you_can_have_progress_without_ gdpled_growth.html, last accessed September 19, 2010.

Lewis, O.T. and M.J.M. Senior. 2011. Assessing conservation status and trends for the world's butterflies: The sampled red list index approach. *Journal of Insect Conservation* 15(1): 121–128.

Li, T.M. 2007. *The will to improve: governmentality, development, and the practice of politics.* Duke University Press, Durham, NC.

Lipschutz, R.D. and J.K. Rowe. 2005. *Globalization, governmentality and global politics: regulation for the rest of us?* Routledge, New York.

Loh, J., R.E. Green, T. Ricketts, J. et al. 2005. The living planet index: using species population time series to track trends in biodiversity. *Philosophical Transactions: Biological Sciences* 360(1454): 289–295.

Lohmann, L. 1995. Visitors to the commons: approaching Thailand's "environmental" struggles from a Western starting point. In *Ecological resistance movements: the global emergence of radical and popular environmentalism,* B.R. Taylor (ed.). State University of New York Press, Albany, pp. 109–126.

Lohmann, L. 2009. Toward a different debate in environmental accounting: the cases of carbon and cost-benefit. *Accounting, Organizations and Society* 34(3–4): 499–534.

Lohmann, L. 2010. Uncertainty markets and carbon markets: variations on Polanyian themes. *New Political Economy* 15(2): 225–254.

Lohmann, L. 2011. The endless algebra of climate markets. *Capitalism, Nature, Socialism* 22(4): 93–116.

Lohmann, L. 2014. Performative equations and neoliberal commodification: the case of climate. In *Nature™ Inc: Environmental Conservation in the Neoliberal Age,* B. Buscher, W. Dressler, and R. Fletcher (eds), pp. 158–181. University of Arizona Press.

Lopa, D., I. Mwanyoka, G. Jambiya, et al. 2012. Towards operational payments for watershed services in Tanzania: a case study from the Uluguru Mountains. *Oryx* 46(1): 34–44.

Lorimer, J. 2007. Nonhuman charisma. *Environment and Planning D: Society and Space* 25(5): 911.

Losurdo, D. 2014. *Liberalism: a counter-history.* Verso Books, London.

Lowe, L. 2015. *Intimacies of four continents.* Duke University Press, Durham, NC.

Luhmann, N. 1989. *Ecological communication.* J. Bednarz Jr. (trans.). University of Chicago Press, Chicago.

Luhmann, N. 1993. *Risk: a sociological theory.* de Gruyter: New York.

Luhmann, N. 2002. *Theories of distinction: re-describing the descriptions of modernity.* W. Rasche (ed.). Stanford University Press, Stanford, CA.

Luke, T.W. 2009. Developing planetarian accountancy: fabricating nature as stock, service, and system for green governmentality. *Current Perspectives in Social Theory* 26: 129–159.

MacArthur, R.H. 1955. Fluctuations of animal populations and a measure of community stability. *Ecology* 36(3): 533–536.

MacArthur, R.H. and E.O. Wilson. 1967. *The theory of island biogeography.* Princeton University Press, Princeton.

MacDonald, K.I. 2010. The devil is in the (bio)diversity: private sector "engagement" and the restructuring of biodiversity conservation. *Antipode* 42(3): 513–550.

MacDonald, K.I. and C. Corson. 2012. "TEEB begins now": a virtual moment in the production of natural capital. *Development and Change* 43: 159–184.

Mace, G., W. Cramer, S. Díaz, et al. 2010. Biodiversity targets after 2010. *Current Opinion in Environmental Sustainability* 2: 1–6.

Mack, J.J. and M. Micacchion. 2006. *An ecological assessment of Ohio mitigation banks: vegetation, amphibians, hydrology and soils.* Ohio EPA technical report WET/2006-1. Ohio Environmental Protection Agency, Division of Surface Water, Wetland Ecology Group, Columbus, OH.

MacKenzie, D.A. 2006. *An engine, not a camera: how financial models shape markets.* MIT Press: Cambridge, MA.

MacKenzie, D.A. 2009. Making things the same: gases, emission rights and the politics of carbon markets. *Accounting, Organizations and Society* 34(3–4): 440–455.

MacKenzie, D.A., F. Muniesa, and L. Siu. 2007. *Do economists make markets?: on the performativity of economics.* Princeton University Press, Princeton.

MacKinnnon, J.B. 2010. A 10 percent world. *Walrus,* September 2010. Available at http://www.walrusmagazine.com/articles/2010.09-environment-a-10-percent-world/, last accessed May 10, 2013.

Madsen, B., N. Carroll, and K. Moore Brand. 2010. *State of biodiversity markets report: offset and compensation programs worldwide.* Washington, DC: Ecosystem Marketplace. Available at http:www.ecosystemmarketplace.com/documents/acrobat/sbdmr.pdf, last accessed September 16, 2014.

Madsen, B., N. Carroll, D. Kandy, and G. Bennett. 2011. Update: state of biodiversity markets. Washington, DC: Forest Trends. Available at http://www.forest-trends.org/documents/files/doc_2848.pdf, last accessed September 16, 2014.

Malua BioBank. 2015. How does it work? Australia: New Forests Asia. Available at http://www.maluabiobank.com/explore.php?id=How_it_works, last accessed February 3, 2015.

Mann, G. 2007. *Our daily bread: wages, workers, and the political economy of the American West.* UNC Press, Chapel Hill.

Mann, G. 2013. *Disassembly required: a field guide to actually existing capitalism*. AK Press, Chico, CA.

Mansfield, B. 2004. Neoliberalism in the oceans: rationalization, property rights, and the commons question. *Geoforum* 35(3): 313–26.

Marris, E. 2011. *Rambunctious garden: saving nature in a post-wild world*. Bloomsbury, New York.

Marsh, G.P. 1864. *Man and nature*. Charles Scribner, New York.

Marvier, M. 2014. A call for ecumenical conservation. *Animal Geography* 17(6): 518–519.

Mason, R. 2011. European carbon market suspended over fraud fears. *Telegraph*, January 19, 2011. Available at http://www.telegraph.co.uk/finance/newsby sector/energy/8269907/European-carbon-market-suspended-over-fraud-fears.html, last accessed February 19, 2015.

Matulis, B.S. 2013. The narrowing gap between vision and execution: neoliberalization of PES in Costa Rica. *Geoforum* 44: 253–260.

Mateo, N., W. Nader, and G. Tamayo. 2001. Bioprospecting. *Encyclopedia of biodiversity, Vol.1*. Academic Press, New York, pp. 471-487.

McAfee, K. 1999. Selling nature to save it? *Society and Space* 17(2): 133–154.

McAfee, K. and E.N. Shapiro. 2010. Payments for ecosystem services in Mexico: nature, neoliberalism, social movements, and the state. *Annals of the Association of American Geographers* 100(3): 579–599.

McCauley, D.J. 2006. Selling out on nature. *Nature* 443(7107): 27–28.

McGraw, D.M. 2002. The CBD – key characteristics and implications for implementation. *Review of European Community & International Environmental Law* 11(1): 17–28.

McKinsey & Company. 2010. *The next environmental issue for business: McKinsey Global survey results*. Available at http://www.mckinsey.com/business-functions/ sustainability-and-resource-productivity/our-insights/the-next-environmental-issue-for-business-mckinsey-global-survey-results,S last accessed March 2, 2016.

McNeely, J.A. 1988. *Economics and biological diversity: developing and using economic incentives to conserve biological resources*. IUCN, Gland, Switzerland.

McNeely, J.A., K.R. Miller, W.V. Reid, R.A. Mittermeier, and T.B. Werner. 1990. *Conserving the world's biological diversity*. IUCN, Gland, Switzerland.

Meadows, D.L., D.H. Meadows, J. Randers, and W.W. Behrens III. 1972. *The limits to growth*. Universe Books, New York.

Mendenhall, C. D., G.C. Daily, and P.R. Ehrlich. 2012. Improving estimates of biodiversity loss. *Biological Conservation* 151(1): 32.

Mendes, R.S., L.R. Evangelista, S.M. Thomaz, et al. 2008. A unified index to measure ecological diversity and species rarity. *Ecography* 31(4): 450–456.

Mies, M. [1986] 1998. *Patriarchy and accumulation on a world scale*. Zed books, London.

Millennium Ecosystem Assessment. 2005. *Ecosystems and human well-being: synthesis*. Island Press, Washington, DC.

Miller, B., M.E. Soulé, and J. Terborgh. 2014. "New conservation" or surrender to development. *Animal Conservation* 17: 509–515.

Mirowski, P. 2013. *Never let a serious crisis go to waste: how neoliberalism survived the financial meltdown*. Verso Books, London.

Mitchell, T. 2002. *Rule of experts: Egypt, techno-politics, modernity*. University of California Press, Berkeley.

Mitchell, T. 2006. The work of economics: how a discipline makes its world. *European Journal of Sociology* 46(2): 297–320.

Mitchell, T. 2007. The properties of markets. In *Do economists make markets?* D. MacKenzie, F. Muniesa, and L. Siu (eds). Princeton University Press, Princeton, pp. 244–275.

Mittermeier, R.A., N. Myers, J.B. Thomsen, et al. 1998. Biodiversity hotspots and major tropical wilderness areas: approaches to setting conservation priorities. *Conservation Biology* 12(3): 516–520.

Monbiot, G. 2009. Monbiot's royal flush: top ten climate deniers. *Guardian*, March 9, 2009. Available at http://www.guardian.co.uk/environ ment/georgemonbiot/2009/mar/06/climate-change-deniers-top-10, last accessed February 3, 2015.

Mooney, H., A. Cropper, and W. Reid. 2005. Confronting the human dilemma: how can ecosystems provide sustainable services to benefit society? *Nature* 434(7033): 561.

Mooney, H.A. and J. Lubchenco. SCOPE Global Biodiversity Assessment Report. UNEP, Nairobi.

Mooney, H.A., J.H. Cushman, E. Medina, O.E. Sala, and E.D. Schulze. 1996. The SCOPE Ecosystem Functioning of Biodiversity Program. In *Functional roles of biodiversity: a global perspective*, H.A. Mooney, J.H. Cushman, E. Medina, O.E. Sala, and E.D. Schulze (eds). John Wiley & Sons, New York, pp. 1–6.

Moore, J.W. 2015. *Capitalism in the web of life: ecology and the accumulation of capital*. Verso Books, London.

Mulder, I. 2007. *Biodiversity, the next challenge for financial institutions?* IUCN, Gland, Switzerland.

Mulder, I. and T. Koellner. 2010. The next challenge for banks? An assessment how banks account for biodiversity risks and opportunities in their business operations. Available from author.

Mullan, K. and T. Swanson. 2009a. *An international market-based instrument to finance biodiversity conservation: towards a green development mechanism*. Report from an expert workshop, Amsterdam, February 9–10, 2009. Available at http://earthmind.org/files/gdm/ReportfromFeb09ExpertWorkshopwithAnnex1. pdf last accessed March 2, 2016.

Mullan, K. and T. Swanson. 2009b. *An international market-based instrument to finance biodiversity conservation: towards a green development mechanism: a proposal for a Green Development Mechanism*. Available at http:// earthmind.org/files/gdm/Proposal_for_a_GDM1.pdf, last accessed March 2, 2016.

Mullan, K., A. Konotolen, and T. Swanson. 2009. *Technical background paper*. Available at http://earthmind.org/files/gdm/Background_Paper1.pdf, last accessed March 2, 2016.

Muniesa, F., Y. Millo, and M. Callon. 2007. An introduction to market devices. *The Sociological Review* 55(s2): 1–12.

Muradian, R., E. Corbera, U. Pascual, et al. 2010. Reconciling theory and practice: an alternative conceptual framework for understanding payments for environmental services. *Ecological Economics* 69(6): 1202–1208.

Murdoch, W., S. Polasky, K.A. Wilson, H.P. Possingham, P. Kareiva, and R. Shaw. 2007. Maximizing return on investment in conservation. *Biological Conservation* 139(3–4): 375–388.

Myers, N. 1979. *The sinking ark: a new look at the problem of disappearing species.* Pergamon Press, Oxford and New York.

Myers, N. 1988a. Tropical forests and their species: going, going. In *Biodiversity*, E.O. Wilson (ed.). National Academy Press, Washington, DC, pp. 28–35.

Myers, N. 1988b. Threatened biotas: "hot spots" in tropical forests. *The Environmentalist* 8(3): 187–208.

Myers, N. 1990. The biodiversity challenge: expanded hot-spots analysis. *The Environmentalist* 10(4): 243–256.

Myers, N. 1993. Tropical Forests: the main deforestation fronts. *Environmental Conservation* 20(1): 9-16.

Myers, N., R.A. Mittermeier, C.G. Mittermeier, et al. 2000. Biodiversity hotspots for conservation priorities. *Nature* 403(6772): 853–858.

Naidoo, R., A. Balmford, P.J. Ferraro, et al. 2006. Integrating economic costs into conservation planning. *Trends in Ecology & Evolution* 21(12): 681–687.

Naidoo, R., A. Balmford, R. Costanza, et al. 2008. Global mapping of ecosystem services and conservation priorities. *Proceedings of the National Academy of Sciences* 105(28): 9495.

Natural Capital Coalition. Undated. Valuing natural capital in business: towards a harmonized protocol. Available http://www.naturalcapitalcoalition.org/js/plugins/filemanager/files/Valuing_Nature_in_Business_Part_1_Framework_WEB.pdf, last accessed February 17, 2015.

Natural Capital Declaration. 2013. The Declaration. Available at: http://www.naturalcapitaldeclaration.org/the-declaration/, last accessed May 8, 2013.

Natural Capital Project. 2011. About page. Available at http://www.naturalcapitalproject.org/about.html, last accessed September 19, 2011.

Natural Capital Project. n.d. Ecosystem Planning in China. http://www.naturalcapitalproject.org/china-case-study/, last accessed March 2, 2016.

Natural Value Initiative. 2008. Dependency and impact on ecosystem services – unmanaged risk, unrealized opportunity: a briefing document for the food, beverage and tobacco sectors. Available at http://www.naturalvalueinitiative.org/uploads/2/7/9/5/27950279/dependency_and_impact_on_ecosystem_services_in_the_food_beverage_and_tabacco_sectors.pdf, last accessed March 15, 2016.

NatureVest and EKO Asset Management Partners. 2014. *Investing in conservation: a landscape assessment of an emerging market.* November 2014. Available at http://www.naturevesttnc.org/pdf/InvestingIn Conservation_Report.pdf, last accessed February 28, 2016.

Nelson, E., G. Mendoza, J. Regetz, et al. 2009. Modelling multiple ecosystem services, biodiversity conservation, commodity production and tradeoffs at landscape scales. *Frontiers in Ecology and Environment* 7(1): 4–11.

Neumann, R.P. 1998. *Imposing wilderness: struggles over livelihood and nature preservation in Africa.* University of California Press, Berkeley.

Norgaard, R.B. 2010. Ecosystem services: from eye-opening metaphor to complexity blinder. *Ecological Economics* 69(6): 1219–1227.

Norse, E.A., K.L. Rosenbaum, D.S. Wilcove et al. 1986. *Conserving biological diversity in our national forests.* The Wilderness Society, Washington DC.

O'Connor, J. 1988. Capitalism, nature, socialism: A theoretical introduction. *Capitalism, Nature, Socialism,* 1(1): 11–38.

O'Connor, J. 1998. *Natural causes: essays in ecological Marxism.* Guilford Press, New York.

Odum, H.T. 1981. *Energy basis for man and nature.* McGraw Hill, New York.

Organisation for Economic Co-operation and Development. 2009. *Summary record of the WGEAB workshop on innovative international financing for biodiversity conservation and sustainable use.* Paris, July 2, 2009. Available at http://www.oecd.org/dataoecd/15/10/44640494.pdf, last accessed September 15, 2011.

Osborn, F. 1948. *Our plundered planet.* Little, Brown and Company, Boston.

Ostfeld, R.S. and K. LoGiudice. 2003. Community disassembly, biodiversity loss, and the erosion of an ecosystem service. *Ecology* 84(6): 1421–1427.

Panayoutou, T. 1995. Conservation of biodiversity and economic development: the concept of transferable development rights. In *Biodiversity conservation,* C.A. Perrings, K.-G. Maler, C. Folke, et al. (eds). Kluwer Academic Publishers, Dordrecht, The Netherlands, pp. 288–303.

Parker, C., M. Cranford, N. Oakes, and M. Leggett. 2012. *The little biodiversity finance book: a guide to proactive investment in natural capital (PINC), 3rd edition.* Oxford, UK: Global Canopy Foundation.

Parry, B. 2004. *Trading the genome: investigating the commodification of bioinformation.* Columbia University Press, New York.

Pearce, D.W. and C.A. Perrings. 1994. Biodiversity conservation and economic development: local and global dimensions. In *Biodiversity conservation,* C.A. Perrings, K.-G. Maler, C. Folke, et al. (eds). Kluwer Academic Publishers, Dordrecht, The Netherlands, pp. 22–39.

Pearce, D.W. and D. Moran. 1994. *The economic value of biodiversity.* IUCN and Earthscan, London.

Pearce, F. 2012. Costing the earth: the value of pricing the planet. October 13, 2012. *New Scientist.*

Peck, J. 2010. *Constructions of neoliberal reason.* Oxford University Press, Oxford.

Peluso, N.L. 2012. What's nature got to do with it? A situated historical perspective on socio-natural commodities. *Development and Change* 43: 79–104.

People's Agreement. 2010. World People's Conference on Climate Change and the Rights of Mother Earth People's Agreement. Available at http://pwccc.wordpress.com/support, last accessed May 3, 2013.

Pereira, H. M. and H. David Cooper. 2006. Towards the global monitoring of biodiversity change. *Trends in Ecology & Evolution* 21(3): 123–129.

Perrings, C.A. 1995a. Preface. In *Biodiversity Loss*, C.A. Perrings, K.-G. Maler, C. Folke, et al. (eds). Cambridge University Press, MA, pp. xi–xii.

Perrings, C.A. 1995b. Preface. In *Biodiversity conservation*, C.A. Perrings, K.-G. Maler, C. Folke, et al. (eds). Kluwer Academic Publishers, Dordrecht, The Netherlands, pp. xv–xvi.

Perrings, C. 1995c. Preface. In Barbier, E., J.C. Burgess, and C. Folke. 1994. *Paradise lost?The ecological economics of biodiversity*. Earthscan, London, pp. x–xv.

Perrings, C.A. 2010. The economics of biodiversity: the evolving agenda. *Environment and Development Economics* 5: 721–746.

Perrings, C.A. and B.W. Walker. 1995. Biodiversity loss and the economics of discontinuous change in semi-arid regions. In *Biodiversity conservation*, C.A. Perrings, K.-G. Maler, C. Folke, et al. (eds). Kluwer Academic Publishers, Dordrecht, The Netherlands, pp. 190–210.

Perrings, C.A. and H. Opschoor. 1994. The loss of biological diversity: some policy implications. *Environmental and Resource Economics* 4(1):1–11.

Perrings, C.A., C. Folke, and K.-G. Maler. 1992. The ecology and economics of biodiversity loss: the research agenda. *Ambio* 21(3): 201–211.

Perrings, C.A., K.-G. Maler, C. Folke, et al. 1995a. Introduction: framing the problem of biodiversity loss. In *Biodiversity loss*, C.A. Perrings, K.-G. Maler, C. Folke, et al. (eds). Cambridge University Press, MA, pp. 1–18.

Perrings, C.A., K.-G. Maler, et al. (eds). 1995b. *Biodiversity loss*. Cambridge University Press, MA.

Perrings, C.A., K.-G. Maler, C. Folke, et al. (eds). 1995c. *Biodiversity conservation*. Kluwer Academic Publishers, Dordrecht, The Netherlands.

Perrings, C.A., K.-G. Maler, C. Folke, et al. 1995d. Biodiversity conservation and economic development: the policy problem. In *Biodiversity conservation*, C.A. Perrings, K.-G. Maler, C. Folke, et al. (eds). Kluwer Academic Publishers, Dordrecht, The Netherlands, pp. 2–21.

Petley, S., J. Grayson, N.M. Gillespie, et al. 2007. Forest-backed bonds proof of concept study. Issue Brief. World Bank Group: IFC Advisory Services in Environmental and Social Sustainability, Available at http://www.ifc.org/wps/wcm/connect/4957218048855503b524f76a6515bb18/IFC_Breif_Forest_web.pdf?MOD=AJPERES&CACHEID=4957218048855503b524f76a6515bb18 last accessed October 3, 2014.

Plumwood, V. 1993. *Feminism and the mastery of nature*. Routledge, London.

Philo, C. 2012. A "new Foucault" with lively implications – or "the crawfish advances sideways." *Transactions of the Institute of British Geographers* 37: 496–514.

Popper, K. 2014. Conjectures and refutations: the growth of scientific knowledge. Routledge, New York.

Porter, T. 1995. *Trust in numbers: the pursuit of objectivity in science and public life*. Princeton University Press, Princeton.

Pratt, G. 2004. *Working feminism*. Edinburgh University Press, Edinburgh.

Province of BC. 2013. Facts about BC's mountain pine beetle. Victoria: Province of British Columbia. Available at http://www.for.gov.bc.ca/hfp/mountain_pine_beetle/Updated-Beetle-Facts_April2013.pdf, last accessed Feb 22, 2015.

Prudham, S. 2003. Taming trees: capital, science, and nature in pacific slope tree improvement. *Annals of the Association of American Geographers* 93(3): 636–656.

Prudham, S. 2007. The fictions of autonomous invention: accumulation by dispossession, commodification and life patents in Canada. *Antipode* 39(3): 406–429.

Quammen, D. 1997. *The song of the dodo: island biogeography in an age of extinctions.* Scribner, New York.

Rabinow, P. 1984. *The Foucault reader.* Pantheon Books, New York.

Raffa, K.F., B.H. Aukema, B.J. Bentz, A.L. Carroll, J.A. Hicke, M.G. Turner, and W.H. Romme. 2008. Cross-scale drivers of natural disturbances prone to anthropogenic amplifications: the dynamics of biome-wide bark beetle eruptions. *Bioscience* 58(6): 501–517.

Rajan, K.S. 2006. *Biocapital: the constitution of postgenomic life.* Duke University Press, Durham, NC.

Randall, A. 1988. What mainstream economists have to say about the value of biodiversity. In *Biodiversity*, E.O. Wilson (ed.). National Academy Press, Washington, DC, pp. 21–26.

Raymond, C.M., G.G. Singh, K. Benessaiah, et al. 2013. Ecosystem services and beyond: using multiple metaphors to understand human–environment relationships. *BioScience:* 63, 536–546.

Rey Benayas, J.M., A.C. Newton, A. Diaz, and J.M. Bullock. 2009. Enhancement of biodiversity and ecosystem services by ecological restoration: A meta-analysis. *Science* 325(5944): 1121–1124.

Redford, K.H. and W.M. Adams. 2009. Payment for ecosystem services and the challenge of saving nature. *Conservation Biology* 23(4): 785–787.

Reid, W.V., H.A. Mooney, D. Capistrano, et al. 2006. Nature: the many benefits of ecosystem services. *Nature* 443(7113): 749–750.

Reid, W.V., S.A. Laird, R. Gámez-Lobo, A. Sittenfeld-Appel, D.H. Janzen, M.A. Gollin, and C. Juma. 1993. *Biodiversity prospecting: using genetic resources for sustainable development.* WRI, Washington.

Reith, G. 2004. Uncertain times: the notion of 'risk' and the development of modernity. *Time and Society* 13(2–3): 383–402.

Ricketts, T.H., G.C. Daily, P.R. Ehrlich, and C.D. Michener. 2004. Economic value of tropical forest to coffee production. *PNAS* 101(34): 12579–12582.

Ridder, B. 2008. Questioning the ecosystem services argument for biodiversity conservation. *Biodiversity and Conservation* 17(4): 781–790.

Robbins, P. and S. Moore. 2013. Ecological anxiety disorder: diagnosing the politics of the anthropocene. *Cultural Geographies* 20(1): 3–19.

Robbins, P. 2014. No going back: the political ethics of ecological novelty. In *Traditional Wisdom and Modern Knowledge for the Earth's Future (International Perspectives in Geography)*, K. Okamoto and Y. Ishikawa (eds). Springer, Tokyo: pp. 103–118.

Rodríguez, J.P., T.D. Beard, E.M. Bennett et al. 2006. Trade-offs across space, time, and ecosystem services. *Ecology and Society* 11(1): Article 28.

Rodríguez, J. P., T. D. Beard, Jr., E. M. Bennett, et al. 2006. Trade-offs across space, time, and ecosystem services. *Ecology and Society* 11(1): 28. Available at http://www.ecologyandsociety.org/vol11/iss1/art28/, last accessed March 2, 2016.

Robertson, M. 2004. The neoliberalization of ecosystem services: wetland mitigation banking and problems in environmental governance. *Geoforum* 35(3): 361–373.

Robertson, M. 2006. The nature that capital can see: science, state, and market in the commodification of ecosystem services. *Environment and Planning D: Society and Space* 24(3): 367–387.

Robertson, M. 2007. Discovering price in all the wrong places: the work of commodity definition and price under neoliberal environmental policy. *Antipode* 39(3): 500–526.

Robertson, M. 2012. Measurement and alienation: making a world of ecosystem services. *Transactions of the Institute for British Geographers* 37: 386–401.

Robertson, M. and N. Hayden. 2008. Evaluation of a market in wetland credits: entrepreneurial wetland banking in Chicago. *Conservation Biology* 22(3): 636–646.

Rockstrom, J., W. Steffen, K. Noone, et al. 2009. A safe operating space for humanity. *Nature* 461(7263): 472–475.

Rose, N. 1999. *Powers of freedom: reframing political thought.* Cambridge University Press, Cambridge.

Roughgarden, J. and F. Smith. 1996. Why fisheries collapse and what to do about it. *PNAS* 93(10): 5078–5083.

Rowe, J.K. 2005. Corporate social responsibility as a business strategy. In *Globalization, governmentality, and global politics: regulation for the rest of us?* R.D. Lipschutz and J.K. Rowe (authors). Routledge, New York, pp. 130–169.

Roy, A. 2010. *Poverty capital: microfinance, and the making of development.* Routledge, New York.

Ruckelshaus, M., E. McKenzie, H. Tallis, et al. 2015. Notes from the field: lessons learned from using ecosystem service approaches to inform real-world decisions. *Ecological Economics* 115(c): 11–21.

Ruhl, J.B. and J.E. Salzman. 2006. The law and policy beginnings of ecosystem services. *Journal of Land Use and Environmental Law* 22(2): 157.

Rutherford, S. 2007. Green governmentality: insights and opportunities in the study of nature's rule. *Progress in Human Geography* 31(3): 291.

Sarkar, S. 1999. Wilderness preservation and biodiversity conservation – keeping divergent goals distinct. *BioScience* 49(5): 405–412.

Sax, D.F., Gaines, S.D. and Brown J.H. 2002. Species invasions exceed extinctions on islands worldwide: a comparative study of plants and birds. *American Naturalist* 160(6): 766–783.

Scholes, R.J., G.M. Mace, W. Turner, et al. 2008. Toward a global biodiversity observing system. *Science* 321(5892): 1044–1045.

Schulze, E.D. and H.A. Mooney (eds). 1993. *Biodiversity and ecosystem function.* Springer, New York.

Scott, J. 1998. *Seeing like a state: how certain schemes to improve the human condition have failed.* Yale University Press, New Haven.

Shapiro-Garza, E. 2013. Contesting market-based conservation: payments for ecosystem services as a surface of engagement for rural social movements in Mexico. *Geoforum* 46: 5–15.

Shiva, V. 1991. Introduction. In *Biodiversity: social and ecological perspectives*, V. Shiva, P. Anderson, H. Schucking, A. Gray, L. Lohmann, and D. Cooper (eds). Zed Books, London and World Rainforest Movement, Montevideo, Uruguay, pp. 7–11.

Shiva, V., J. Bandyopadhyay, P. Hegde, et al. 1991. Ecology and the politics of survival: conflicts over natural resources in India. Sage, New Delhi.

Shiva, V. 1997. *Biopiracy: the plunder of nature and knowledge.* South End Press, New York.

Shiva, V., P. Anderson, H. Schucking, A. Gray, L. Lohmann, and D. Cooper. 1991. *Biodiversity: social and ecological perspectives.* Zed Books, London and World Rainforest Movement, Montevideo, Uruguay.

Shiva, V. 2012. Rio+20: an undesirable turn. *Common Dreams* July 3 2012. Available at http://www.commondreams.org/views/2012/07/03/rio20-undesirable-u-turn, last accessed October 9, 2015

Shiva, V. 2014. We Are the Soil. *The Asian Age*, May 24, 2014. Available at : http://www.commondreams.org/views/2014/05/26/we-are-soil, last accessed February 27, 2016.

Simonit, S. and C. Perrings. 2013. Bundling ecosystem services in the Panama Canal Watershed. *PNAS* 110(23).

Sklair, L. 2001. *The transnational capitalist class.* Wiley-Blackwell, Oxford.

Smith, N. 1984. *Uneven development: nature, capital and the production of space.* University of Georgia Press, Athens, Georgia.

Smith, N. 2007. Nature as accumulation strategy. *Socialist Register* 43:16–36.

Soulé, M. 2013. The "new conservation." *Conservation Biology* 27(5): 895–897.

Specter, M. 2014. Seeds of doubt: An activist's controversial crusade against genetically modified crops. *New Yorker,* August 25, 2014, Available at http://www.newyorker.com/magazine/2014/08/25/seeds-of-doubt, last accessed February 27, 2016.

Spurgeon, J.P.G. 2014. B@B workstream 1: natural capital accounting for business: guide to selecting an approach. Available at http://ec.europa.eu/environment/biodiversity/business/assets/pdf/b-at-b-platform-nca-workstream-final-report.pdf, last accessed February 17, 2015.

Stabroek News. 2012. Iwokrama, Canopy Capital forest deal collapses. February 19, 2012. Available at http://www.stabroeknews.com/2012/news/stories/02/19/iwokrama-canopy-capital-forest-deal-collapses/, last accessed February 28, 2016.

Star, S. and J. Griesemer, J. 1989. Institutional ecology, translations, and boundary objects. *Social Studies of Science* 19(3): 387–420.

Starzomski, B. 2013. Novel ecosystems and climate change. In R.J. Hobbs, E. Higgs, and C. Hall (eds). *Novel ecosystems: intervening in the new ecological world order*. John Wiley & Sons, Oxford. pp. 88–101.

Stepan, N.L. 2006. *Picturing tropical nature*. Reaktion Books, London.

Stern, N. 2006. *The Stern review on the economics of climate change*. HM Government, London. Available at http://mudancasclimaticas.cptec.inpe.br/~rmclima/pdfs/destaques/sternreview_report_complete.pdf, last accessed April 20, 2016.

Stoler, A. 2008. Imperial debris: reflections on ruin and ruination. *Cultural Anthropology* 23(2): 191–219.

Stoler, A. 2013. *Imperial debris: on ruins and ruination*. Duke University Press, Durham, NC.

Sturgeon, N. 1997. *Ecofeminist natures: race, gender, feminist theory and political action*. Routledge, New York.

Sullivan, S. 2010a. "Ecosystem service commodities" – a new imperial ecology? Implications for animist immanent ecologies, with Deleuze and Guattari. *New Formations* 69 (Spring 2010): 111–28.

Sullivan, S. 2010b. The environmentality of "Earth Incorporated": on contemporary primitive accumulation and the financialisation of environmental conservation. Paper presented at the conference An Environmental History of Neoliberalism, Lund University, May 6–8, 2010.

Sullivan, S. 2013. Banking nature? The spectacular financialisation of environmental conservation. *Antipode* 45(1): 198–217.

Sundberg, J. 2006. Conservation encounters: transculturation in the "contact zones" of empire. *Cultural Geographies* 13(2): 239–265.

Sundberg, J. 2011. Diabolic *caminos* in the desert and cat fights on the Rio: a posthumanist political ecology of boundary enforcement in the United States–Mexico borderlands. *Annals of the Association of American Geographers* 101(2): 318–336.

Sundberg, J. and J. Dempsey. 2009. Culture/Nature. In *International Encyclopedia of Human Geography*, R. Kitchin and N. Thrift (eds.), Elsevier Press, Oxford.

Swanson, T. 1995. The international regulation of biodiversity decline: optimal policy and evolutionary product. In *Biodiversity loss*, C.A. Perrings, K.-G. Maler, C. Folke, et al. (eds). Cambridge University Press, MA, pp. 225–259.

Takacs, D. 1996. *The idea of biodiversity*. Johns Hopkins University Press, Baltimore.

Tallis, H. and P. Kareiva. 2006. Ecosystem Services. *Current Biology* 15(17): 746–748.

Tallis, H. and S. Polasky. 2009. Mapping and valuing ecosystem services as an approach for conservation and natural resource management. *Annals of the New York Academy of Sciences* 1162, 265–283.

Tallis, H.T., T. Ricketts, A.D. Guerry, et al. 2013. InVEST 2.5.3 User's Guide. The Natural Capital Project, Stanford, Available at http://ncp-dev.stanford.edu/~dataportal/invest-releases/documentation/current_release/, last accessed February 22, 2015.

Taylor, C. 1995. *Two theories of modernity.* The Hastings Center Report 25(2): 24–33.

The Economics of Ecosystems and Biodiversity. 2008. *The economics of ecosystems and biodiversity: An interim report.* European Commission and German Federal Ministry for the Environment, Cambridge, UK.

The Economics of Ecosystems and Biodiversity. 2009. *TEEB for national and international policy makers – summary: responding to the value of nature.* Available at http://www.teebweb.org/media/2009/11/National-Executive-Summary_-English.pdf, last accessed March 2, 2016.

The Economics of Ecosystems and Biodiversity. 2010a. *TEEB report for business – executive summary.* Available at http://img.teebweb.org/wp-content/uploads/Study%20and%20Reports/Reports/Business%20and%20 Enterprise/Executive%20Summary/Business%20Executive%20Summary_ English.pdf, last accessed March 2, 2016.

The Economics of Ecosystems and Biodiversity. 2010b. *The economics of ecosystems and biodiversity: the ecological and economic foundations.* Earthscan. Available at http://www.teebweb.org/our-publications/teeb-study-reports/ ecological-and-economic-foundations/, last accessed March 2, 2016.

Third World Network. 2010. *Mixed reactions on new access and benefit sharing treaty.* Available at http://www.twn.my/title2/resurgence/2010/242-243/cover02. htm, last accessed March 2, 2016.

Third World Network. n.d. *The politics of knowledge at the CBD.* Available at http://www.twn.my/title2/resurgence/2010/242-243/cover02.htm, last accessed March 2, 2016.

Thompson, R. and B.M. Starzomski. 2007. What does biodiversity actually do? A review for managers and policy makers. *Biodiversity and Conservation* 16(5): 1359–1378.

Tinker, C. 1995. A "new breed" of treaty: The United Nations Convention on Biological Diversity. *Pace Environmental Law Review* 12(2): 191–222.

Trucost. 2013. *Natural capital at risk.* Available at http://www.naturalcapi talcoalition.org/projects/natural-capital-at-risk.html, last accessed March 2, 2016.

Tsing, A.L. 2005. *Friction: an ethnography of global connection.* Princeton University Press, Princeton.

Tuomisto, H. 2012. An updated consumer's guide to evenness and related indices. *Oikos*, 121(8): 1203–1218.

Turner, M. 2014. Political ecology I: an alliance with resilience. *Progress in Human Geography* 38: 616–623.

Turner, R.K., C. Folke, I.M. Green, and I.J. Bateman. 1995. Wetland valuation: three case studies. In *Biodiversity conservation*, C.A. Perrings, K.-G. Maler, C. Folke, et al. (eds). Kluwer Academic Publishers, Dordrecht, The Netherlands, pp. 129–149.

UNDP. 2010. Biodiversity Superpower website. Available at http://www.latina merica.undp.org/content/rblac/en/home/presscenter/pressreleases/2010/02/ amrica-latina-y-el-caribe-superpotencias-de-biodiversidad.html last accessed March 2, 2016.

UNEP. 2010. Where on earth is biodiversity? Available at http://www.unep.org/ Documents.Multilingual/Default.asp?DocumentID=646&ArticleID=6695&l=en, last accessed October 7, 2010.

UNEP FI. 2007. Summary Outcomes. First Group Workshop of UNEP FI's Biodiversity & Ecosystem Services Workstream, London, 5 April 2007. Available at http://www.unepfi.org/fileadmin/events/2007/london/outcome_ summary.pdf, last accessed May 2, 2013.

UNEP FI. 2009. *The global state of sustainable insurance: understanding and integrating environmental, social and governance factors in insurance.* UNEP FI, Geneva.

UNEP FI. 2010. *CEO briefing: Demystifying materiality.* UNEP FI, Geneva.

UNEP FI's Biodiversity & Ecosystem Services Workstream. 2007. Notes from meeting, April 5, 2007, London. Available at http://www.unepfi.org/filead min/events/2007/london/outcome_summary.pdf, last accessed September 11, 2011.

UNEP FI. 2011. REDDy SET GROW. Biodiversity and Ecosystems Workstream and Climate Change Working Group, UNEP Finance Initiative. Geneva: UNEP FI. Available at http://www.unepfi.org/fileadmin/documents/reddysetgrow. pdf, last accessed February 2, 2015.

UNEP FI. 2012. A new angle on sovereign credit risk: environmental risk integration in sovereign credit analysis. Available at http://www.unep.org/ PDF/PressReleases/UNEP_ERISC_Final_LowRes.pdf, last accessed February 17, 2015.

UN-REDD. 2013. About the UN-REDD Programme. Available at: http://www. un-redd.org, last accessed May 2, 2013.

VFU Forum. 2011. Biodiversity principles: recommendations for the financial sector. Available at http://www.unepfi.org/fileadmin/documents/biodiversity_ principles_en.pdf, last accessed February 17, 2015.

Van Paddenburg, A., A. Bassi, E. Buter, et al. 2012. *Heart of Borneo: investing in nature for a green economy.* WWF Heart of Borneo Global Initiative, Jakarta.

Vidal, J. 2012. Cut world population and redistribute resources, expert urges. *Guardian.* April 6, 2012. Available at http://www.theguardian.com/envi ronment/2012/apr/26/world-population-resources-paul-ehrlich, last accessed Feb 22, 2015.

Vogt, W. 1948. *Road to survival.* William Sloan, New York.

Voosen, P. 2011. Scientists clash on claims over extinction "overestimates." *New York Times,* May 18, 2011. Available at http://www.nytimes.com/ gwire/2011/05/18/18greenwire-scientists-clash-on-claims-over-extinction- ove-96307.html?pagewanted=all, last accessed September 16, 2011.

Vorhies, F. 2009. Comments of Francis Vorhies on the Amsterdam Meeting. Available from author.

Walker B.H. 1992. Biological diversity and ecological redundancy. *Conservation Biology,* 6(1): 18–23.

Walker, B.H. 1994. Rangeland ecology: managing change in biodiversity. In *Biodiversity conservation,* C.A. Perrings, K.-G. Maler, C. Folke, C.S. Holling, and B.-O. Jansson (eds). Kluwer Academic Publishers, Dordrecht, The Netherlands, pp. 65–81.

Walker, J. and M. Cooper. 2011 Genealogies of resilience: from systems ecology to the political economy of crisis adaptation. *Security Dialogue* 42(2):143–160.

Walker, S., A.L. Brower, R.T. Theo Stephens, and W.G. Lee. 2009. Why bartering biodiversity fails. *Conservation Letters* 2(4): 149–157.

Walsh, B. 2013. If carbon markets can't work in Europe, can they work anywhere? *TIME*. Available at http://science.time.com/2013/04/17/if-carbon-markets-cant-work-in-europe-can-they-work-anywhere/, last accessed May 3, 2013.

Wara, M. 2007. Is the global carbon market working? *Nature* 445(8): 595–596.

Watson, J.E.M. et al. 2014. The performance and potential of protected areas. *Nature* 515: 67–73.

WeADAPT, 2012. Equitable payments for watershed services (EPWS) in the Uluguru Mountains, Tanzania. Available at http://weadapt.org/knowledge-base/ecosystem-based-adaptation/the-equitable-payments-for-watershed-services-epws-project-in-the-uluguru-mountains-tanzania, last accessed October 3, 2014.

Weitzman, M.L. 1993. What to preserve? An application of diversity theory to crane conservation. *The Quarterly Journal of Economics* 108(1): 157–183.

Weitzman, M.L. 1995. Diversity functions. In *Biodiversity loss*, C.A. Perrings, K.-G. Maler, C. Folke, et al. (eds). Cambridge University Press, MA, pp. 21–43.

Whatmore, S. 2006. Materialist returns: practising cultural geography in and for a more-than-human world. *Cultural Geographies* 13(4): 600.

West, P. 2006. *Conservation is our government now*. Duke University Press, Durham, NC.

Wilson, E.O. 1984. *Biophilia*. Harvard University Press, Cambridge.

Wilson, E.O. 1988a. Editor's foreword. In *Biodiversity*, E.O. Wilson (ed.). National Academy Press, Washington, DC, pp. v–vii.

Wilson, E.O. 1988b. The current state of biodiversity. In *Biodiversity*, E.O. Wilson (ed.). National Academy Press, Washington, DC, pp. 3–18.

World Business Council for Sustainable Development. n.d. *Guide to corporate ecosystem valuation: a framework for improving corporate decision-making*. Available at http://www.wbcsd.org/work-program/ecosystems/cev/downloads.aspx, last accessed February 17, 2015.

World Business Council on Sustainable Development 2010a. *Post-2010 biodiversity targets: a business perspective*. Available at http://www.wbcsd.org/Pages/EDocument/EDocumentDetails.aspx?ID=12770&NoSearchContextKey=true, last accessed March 2, 2016.

World Business Council on Sustainable Development. 2010b. *Vision 2050: the new agenda for business*. WBCSD, Geneva.

World Commission on Environment and Development (WCED). 1987. *Our common future*. Oxford University Press, Oxford.

World Economic Forum. 2010a. *Global Risks 2010*. WEF, Geneva.

World Economic Forum. 2010b. *Biodiversity and business risk*. WEF, Geneva.

World Resources Institute. 2013. *The corporate ecosystem services review: guidelines for identifying business risks and opportunities arising from ecosystem change*. Available at http://www.wri.org/publication/corporate-ecosystem-services-review, last accessed May 6, 2013.

Wuethner et al. 2014. *Keeping the wild*. Island Press, Washington, DC.

Youatt, R. 2008. Counting species: biopower and the global biodiversity census. *Environmental Values* 17(3): 393–417.

Index

Enterprising Nature: Economics, Markets, and Finance in Global Biodiversity Politics,
First Edition. Jessica Dempsey.
© 2016 John Wiley & Sons, Ltd. Published 2016 by John Wiley & Sons, Ltd.

Printed and bound by CPI Group (UK) Ltd, Croydon, CR0 4YY

27/10/2024

14580298-0001